# A Unifying Theory of Evolution

## generated by means of information modelling

# Studies in Health Technology and Informatics

This book series was started in 1990 to promote research conducted under the auspices of the EC programmes' Advanced Informatics in Medicine (AIM) and Biomedical and Health Research (BHR) bioengineering branch. A driving aspect of international health informatics is that telecommunication technology, rehabilitative technology, intelligent home technology and many other components are moving together and form one integrated world of information and communication media. The series has been accepted by MEDLINE/PubMed, SciVerse Scopus, EMCare, Book Citation Index – Science and Thomson Reuters' Conference Proceedings Citation Index.



## Volume 230

*Recently published in this series*

Vol. 229. H. Petrie, J. Darzentas, T. Walsh, D. Swallow, L. Sandoval, A. Lewis and C. Power (Eds.), Universal Design 2016: Learning from the Past, Designing for the Future – Proceedings of the 3rd International Conference on Universal Design (UD 2016), York, United Kingdom, August 21–24, 2016

Vol. 228. A. Hoerbst, W.O. Hackl, N. de Keizer, H.-U. Prokosch, M. Hercigonja-Szekeres and S. de Lusignan (Eds.), Exploring Complexity in Health: An Interdisciplinary Systems Approach – Proceedings of MIE2016 at HEC2016

Vol. 227. A. Georgiou, L.K. Schaper and S. Whetton (Eds.), Digital Health Innovation for Consumers, Clinicians, Connectivity and Community – Selected Papers from the 24th Australian National Health Informatics Conference (HIC 2016)

Vol. 226. J. Mantas, A. Hasman, P. Gallos, A. Kolokathi and M.S. Househ (Eds.), Unifying the Applications and Foundations of Biomedical and Health Informatics

Vol. 225. W. Sermeus, P.M. Procter and P. Weber (Eds.), Nursing Informatics 2016 – eHealth for All: Every Level Collaboration – From Project to Realization

Vol. 224. N. Maglaveras and E. Gizeli (Eds.), pHealth 2016 – Proceedings of the 13th International Conference on Wearable Micro and Nano Technologies for Personalised Health, 29–31 May 2016, Heraklion, Crete, Greece

Vol. 223. G. Schreier, E. Ammenwerth, A. Hörbst and D. Hayn (Eds.), Health Informatics Meets eHealth – Predictive Modeling in Healthcare – From Prediction to Prevention – Proceedings of the 10th eHealth2016 Conference

ISSN 0926-9630 (print)
ISSN 1879-8365 (online)

# A Unifying Theory of Evolution

## generated by means of information modelling

Jytte BRENDER McNAIR

*Emeritus assoc. professor*
*Dept. of Health Science and Technology*
*Aalborg University*
*Aalborg, Denmark*

ISBN 978-1-61499-687-3 (print)
ISBN 978-1-61499-688-0 (online)
Library of Congress Control Number: 2016949861
doi:10.3233/978-1-61499-688-0-i

Cover image by Lynnclaire Dennis

*Publisher*
IOS Press BV
Nieuwe Hemweg 6B
1013 BG Amsterdam
Netherlands
fax: +31 20 687 0019
e-mail: order@iospress.nl

*Distributor in the USA and Canada*
IOS Press, Inc.
4502 Rachael Manor Drive
Fairfax, VA 22032
USA
fax: +1 703 323 3668
e-mail: iosbooks@iospress.com



PRINTED IN THE NETHERLANDS

DEDICATION

… to my sister in heart and spirit, Lynnclaire Dennis.

A small notice in a magazine in 1999 told about an American woman, who died over the Alps in a hot-air balloon race in 1987. I remembered the headlines in the news when it happened. The essence of the story was that she saw a geometric structure while she was clinically dead. There was a link to a web page, and on that there was a note of a book telling the story, but no information on the price of the book. I pushed the 'contact' button and sent an email asking the publisher "What is the price…" of Lynnclaire's book – and nothing more – and that email was the entry to now 16 years of cooperation with Lynnclaire Dennis: That day, quite unusually, it was Lynnclaire answering incoming emails regarding the book. In less than four months she was at the doorstep of my husband and me, and another four months later she moved permanently from the USA to our house and stayed for 33 months, when she got her own apartment in Switzerland.

Living with Lynnclaire in our house was like living on a volcano; nothing ever became the same again, and numerous unbelievable events and synchronicities popped up every day. She saved my sanity more than once. It seems as if the scientific discoveries on the Mereon Matrix escalated the minute she was permanently in our house; the scientists modelled the geometric structure, the topology and the dynamics of Mereon according to Lynnclaire's memory of what she saw when she died the first time. Now, as Lou Kauffman, professor of mathematics, University of Illinois, Chicago, expressed it in our book on Mereon: "For whatever Mereon really is, she is our link with the creative process at the heart of the universe.". Over the years, I have had so many dreams and messages in simple meditations as well as synchronicities in my everyday life, pointing me in this direction, so that there cannot be any doubt left. This structure is what Lynnclaire brought back from three near-death events, and which Nechung Kuten called "a gift to mankind".

Lynnclaire is incredibly gifted in countless and remarkable ways, self-taught, as an artist, as an author, a teacher, a social architect, and much more; and it is incredible how fast she comprehends and applies the mind-set of a scientist once she is told how. Most remarkably are her social skills, her ability to relate to people and communicate with them, and her ability to make the links between the geometric and topological dynamics of Mereon and 'real-world' situations.

It is an honour to know her, and even more so to be part of the endeavour of exploring the Mereon Matrix from a hard-core scientific perspective.

# Preface

The present study is primarily aimed at an audience of evolutionary biologists in the broad sense for Parts 1-3, as well as at scientists from other professional disciplines looking for new approaches to explore phenomena in nature. However, the content is in general not written as technical text or filled with equations that make it utterly incomprehensible for 'normal' people, although they would have to skip strictly professional biochemical and biological terms that are a constituent part of the evidence in a scientific sense. Therefore, large parts of Part 2 may also be enjoyed by a broader audience with little prior insight into evolutionary biology or science, when in search of explanations for the phenomena that we see in nature.

Finally, Part 4 – with a view into Parts 1-3 – is meant for systems analysts and researchers striving to find a general systems theory. Parts 3 and 4 are the only parts that are not relevant for a general audience.

*A Note for the Reader*

Extensive use of references in scientific literature is generally perceived as evidence in favour of an author's statement when synthesizing findings in the literature on a given topic. At the same time, when applying citations of such literature one has to copy the text literally including printing errors, references and everything relevant, unless otherwise explicitly stated. Including such references in cited text often a) creates confusion because such references are not included in the reference list even if they appear in the text; and furthermore b) they tend to add an unnecessary cognitive burden on the reader. Therefore, references with cited text in the present publication have been replaced by the Italicized text '*ref*' or '*refs*', i.e. single or plural, depending on the number of references omitted in a given place.

*About the Author/Contributors*

The author, JBMcN, is the initiator and sole contributor to the present study. JBMcN and her husband, Dr. Peter McNair, in close cooperation developed the ATCG model described in [1] ('Applied Theory accounting for human moleCular Genetics'), which constitutes a solid foundation for the present work. The author has a master degree in biochemistry, a master degree in computer science (systems development, informatics) and a European Doctorate and PhD in health informatics. The author has had a full-time research position for about 35 years, divided on a university hospital, the software industry and a university, while cross-fertilising the two professions; the time in the industry and the university mainly was dedicated to participation in large and advanced EU R&D Research projects. This combination of professional experience has enabled the present study.

The present study was performed outside of any job or grant affiliation, while the author had an emeritus position at the Aalborg University. The author has since the year 2000 been a member of the scientific team describing the Mereon Matrix (for this, see [2]), went on early retirement in 2011 to work full time on the Mereon Matrix, and is independently continuing such scientific investigations on the matrix.

Contact information: jytte@brender.dk and jbr@hst.aau.dk.

*Acknowledgements*

The author is deeply grateful for her position as Emeritus Assoc. Professor at her former university affiliation at Aalborg University, Dept. of Health Science and Technology, as this enabled a full and free access to most of the original scientific literature. Without this, the present study would have been critically hampered.

The author is deeply grateful for continuous brainstorming and constructive feedback on and in the process with her husband, Dr. Peter McNair, retired director of the Medical Genetics Laboratory, KennedyCenter (DK-2600 Glostrup, Denmark).

*Conflicting Interests*

The author declares that there was and is no conflicting interests, financial or otherwise, that would or could have influenced the outcome of the study.

The author's work on the Mereon Matrix was not financially supported by any external source of funds, except for 1) a couple of travels in 2012, funded by The Mereon Legacy CIC[1] (see www.mereon.org), with the purpose of enabling the team of authors and editors to meet face-to-face during the process of writing the book, [2], and 2) sponsoring the publication expenses for the present book, thereby facilitating its availability as an Open Access publication.

[1] CIC is an acronym for 'Community Interest Company', a non-profit organisation located in the United Kingdom.

# Contents

# Foreword

> "To make progress, scientists must specify phenomena that require explanation, identify causes and decide on what methods, data and analyses are explanatorily sufficient. In doing so, they may inadvertently create a 'conceptual framework'—a way of thinking for their field, with associated assumptions, concepts, rules and practice, that allows them to get on with their work [*refs*]. Conceptual frameworks are necessary in science, but they, and their associated practices, inevitably encourage some lines of research more readily than others. Hence, it is vital that the conceptual frameworks themselves evolve in response to new data, theories and methodologies. This is not always straightforward, as habits of thought and practice are often deeply entrenched. In this regard, alternative conceptual frameworks can be valuable because they draw attention to constructive new ways of thinking, additional causal influences, alternative predictions or new lines of enquiry."
>
> (Citation from [3], with permission)

This citation implicitly nails the scenario of introducing a new perspective on a research topic into an already established scientific community, touching on why controversial new theories tend to create resistance. This has happened many times in science. A couple of well-known examples are a) the shift from Earth being centre of the Universe to a Heliocentric view, b) Darwin's Origin of Species. A couple of recent examples are: c) prions, and d) that peptic ulcer is caused by a bacterial infection. In all of these cases, when these new theories came out, they were forcefully rejected, if not ridiculed for an extended period, by colleagues in the scientific community – and beyond – because these theories were too controversial and contrasting existing / existential belief systems. However, over time these new theories were gradually proven to be valid and finally accepted. This is presently happening for the Extended Evolutionary Synthesis' perspective on evolutionary biology, and for instance also for the Constructive Neutral Evolution theory, and may happen for the theory in the present book as well. Time will show.

One of the arguments may be that the author of the present book has no a priori scientific background within the domain of biological evolution and opponents may therefore ask "why publish a book on evolution, when it is not the author's core competence?" This may for some constitute a basis for critique. However, a combined background in biochemistry and informatics (information modelling), more than three decades of research experience and two years of focussed effort (which is more than most PhD students spend on familiarising with a topic before applying it in practise) should justify writing a kind of review like the present.

The author's professional background is biochemistry and computer science (informatics / information science), and for decades she has bridged these two disciplines in various ways and areas, mainly in large EU R&D projects, and mainly through information modelling in biochemistry and systems development. This ranges from aspects of quality of the semantics of medical knowledge, over the influence of cultural aspects on systems development, to factors in laboratory practise in clinical biochemistry and in diagnostic genetics. Her specialisation within information modelling has been on quality management and constructive evaluation at change processes. So, a very wide background, but all along in some way or another, these research activities have been focused on converting segregated information into models as representations that will facilitate an understanding within a context. So is this book.

The last decade, the focus has increasingly been on application of a particular template information model, and the present study was initiated to assess the general applicability of that template information model for a complex knowledge domain. The original choice regarding an application domain was molecular human genetics, and this modelling endeavour was successfully accomplished and is presented in [1]. The next step was to pick a sub-topic from within that model for a more detailed analysis – again by information modelling – and by chance the choice was the topic of biological evolution. What was intended as a small report became a book, because the topic was so captivating and the modelling rewarding. It was the Nature paper of Laland and co-workers ([4]) that during the modelling effort triggered the inclusion of perspectives from the template information model into the modelling work and led to a framework that in a seamless way unifies all large and small evolutionary theories and puts them into place in the shared framework presented in this book.

The reason that the topic was so captivating was that while the initial investigation of the knowledge domain revealed a chaotic scenery of theories, small and large, of which some seemed to be furiously competing (although not necessarily conflicting), the template information model provided the instrument for reconciliation. The result was a framework, which unifies the Standard Evolutionary Theory with the Extended Evolutionary Synthesis and the Inevitable Evolution theory into one evolutionary theory. Moreover, the template information model pointed at missing topics, resulting in renewed literature searches, and thereby gradually the entire framework was filled with arguments and examples.

The purpose of the present modelling effort was and continues to be the continued investigation of the capability of the Mereon Matrix template information model to serve as a universal system, delimited to the specific sub-domain of biological evolution. Consequently, the modelling efforts will continue with other knowledge domains from within the domain of molecular genetics.

# Part 1: Basics First

Part 1 brings the information necessary for others to be able to repeat the present study or repeat the study for another knowledge domain. Since the book deliberately separates material for the two target audiences (evolutionary biologists and systems analysts), and in case someone wants to perform a similar study, the methodological approach in Part 1 (dedicated for evolutionary biologists) has to be read together with the similar one in Part 4 (dedicated for the domain of informatics / systems analysts/ information modelling).

# 1 Introduction

**Abstract.** The purpose of this chapter is to bring the reader up to date with the motivation, the mind-set and the rationale of the study/work covered and the outcome described, - with relevant references to literature and/or other material that puts the study into a perspective.

The purpose of the present study is to continue investigating the capability of the Mereon Matrix to serve as a generally applicable template information model, through elaboration of the model of human molecular genetics in [1], however at present delimited to a specific sub-domain, evolution.

**Keywords.** Mereon Matrix, template information model, evolution

Modelling a system or phenomena means to capture an abstract representation of its very essence. While the author perceives a theory as "a set of hypotheses related by logical or mathematical arguments to explain and predict a wide variety of connected phenomena in general terms. …", [5], a model is a simplified representation of an understanding of a phenomenon or system. Then, a template model is the means derived from such theory that enables one to apply the theory in practice for particular purposes, for instance to gain a model of a system's phenomenological behaviour. One may draw an analogy between a template information model and the syntax of a language, as the template consists of component parts and rules that tie them together in a strict order, and into which one may fill 'words' to describe the intended meaning.

Modelling, qualitative (phenomenological) as well as quantitative (computational, mathematical or statistical, e.g. for simulation), of systems has[2] been used in science for centuries to achieve a better understanding of a system or a phenomenon. It is an important instrument in scientific work to achieve new insight or assert an existing understanding and in particular as a means for predictive purposes. Physicists, mathematicians and many more seek to find a universal model, one theory that includes everything, a 'Theory of Everything', and which will connect all aspects of physics and bridge natural science with humanities; a giant step. Austrian biologist (and founder of the domain of General System Theory, GST) Ludwig von Bertalanffy was one such individual; his quest was to find a universal template for systems. The following quote underscores his drive:

> "Thus, there exist models, principles, and laws that apply to generalized systems or their subclasses, irrespective of their particular kind, the nature of their component elements, and the relationships or "forces" between them. It seems legitimate to ask for a theory, not of systems of a more or less special kind, but of universal principles applying to systems in general."
>
> ([6] *page 32*)

---

[2] A system is defined as "An organisation in which all structural components and dynamics are interrelational, participating internally, and affecting conditions externally" [2] (*page 480*).

Two additional citations discussing modelling of biological systems shall be emphasized here:

> "... to explain these developments in terms of the properties of cell and developmental systems will unify biology into a set of common principles that can be applied to different systems ...",

and

> "... if we think what are the properties of those systems, and study the mechanisms that are being employed and how they're being modified to achieve their physiological function, we'll have a better understanding of those systems, ...."
>
> (both from [7] *page 6*)

What Kirschner in reality says is that modelling of biological systems may provide the answer to von Bertalanffy's request for a General Systems Theory, and that this may provide a better chance of modifying them and even understanding the pathology of the systems.

von Bertalanffy's idea of a general system theory is adopted. While the author is convinced that the Mereon Matrix's information model (see [2]) constitutes a template for such a general systems theory, it is necessary to work one's way one step at a time toward the long-term goal of achieving a complete systems theory based on the Mereon Matrix's information model.

The concept of the Mereon Matrix has been the subject of a hardnosed epistemological analysis and extensive computer simulations for 20++ years, by physicists, mathematicians and more, and is described exhaustively in [2], including the Mereon Matrix's template information model. The original modelling may be found at http://www.rwgrayprojects.com/Lynn/Lynn01.html. The biological topic that the present book covers has been analysed in terms of this knowledge.

Then, what is Mereon? Professor Louis Kauffman from The University of Chicago, Illinois, expressed it in the book on Mereon, "For whatever Mereon really is, she is our link with the creative process at the heart of the universe." ([2] *page xxxix*). A short summary of the Mereon Matrix is included in Part 4, comprising a description of relevant parts and aspects to a point necessary for grasping the modelling aspects behind the present book. The purpose of the present study is to continue investigating the capability of the Mereon Matrix to serve as a generally applicable template information model, through elaboration of the model of human molecular genetics in [1], however at present delimited to a specific sub-domain, evolution.

All of this was the starting point and motivation for launching the present study.

Part 2 briefly describes individual evolutionary theories when placed in the Unifying Theory, and Part 3 discusses the validity of the outcome of the study and the template information model.

In Part 4, details on the modelling itself are presented and thereby the reasoning behind how the template information model works, providing an account of how it may be applied to real problems, and why the different evolutionary theories fit into the template information model of the Mereon Matrix. Also briefly introduced in this chapter is the Mereon Matrix itself.

### *1.1 Background*

A number of overview/reviews of evolutionary strategies reveals diverse perspectives on the pile of evolutionary theories and elements thereof. Theories on evolution within the literature appear competitive, authors fighting to be right or even just to be heard. One explanation might be that the domain still lacks a completely unified theory of evolution that can fully accommodate all of the complexities of the evolutionary processes. *This is precisely what the present book is about*, while it is NOT the purpose to modify, extend or reconcile any of the existing theories; the purpose IS solely their unification.

The hypothesis is – and subsequently attempted to verify in real application to disprove – the value of the template information model as a generally applicable approach for modelling systems, thus seeking to answer von Bertalanffy's call for a General Systems Theory. If the hypothesis is accurate, then the template model has to be applicable for any knowledge domain, from domains like physics, biology, and medicine, to organisational and social sciences including psychology and more. No single individual can do all of this, and the task is so huge that it has to be divided into smaller tasks for a series of the individual professional disciplines. The template information model was published only in 2013, and thus the team of investigators is only in the beginning of such endeavour. The approach is to carry out a trial, practical application at increasing detail and for a highly complex knowledge domain. Therefore, our pilot study as well as the present study both have a focus on the domain of molecular genetics, because this is extremely complex, yet feasible to address for a researcher educated as a biochemist, while the systems modelling is enabled by the author's background and experience as a systems analyst/informatician.

What does it mean to be valid as a generally applicable template for information modelling? That it will be feasible to model a complete system by means of existing knowledge within the application domain and bring sense to it within the new structure, nothing superfluous, nothing missing. The system of the pilot study was human molecular genetics; in the present study the system is biological evolution – broadening the application range from the pilot study to include all species, simply because evolution of the human species is too complicated and not yet sufficiently mature as a knowledge domain.

This work has been accused by a reviewer of being related to the New Age movement, presumably because the Mereon Matrix first appeared in a near-death-experience. Such accusation demands a comment. Similar events and contributions has happened to other highly esteemed scientists like Buckminster-Fuller and more. To the author, there is a huge difference between the New Age movement and a scientific study in that the former is founded on a belief system with little or weakly objective and measurable facts that may lead society toward solid evidence and hence it belongs in the category of religions. In contrast, a scientific study is founded on a set of stringent rules and principles that ensures a stepwise and progressive accumulation of unbiased evidence. The foundation behind the present study is more than a decade of simulation studies by means of a computational model of the geometry, topology and the dynamics that gradually evolved. All of this is documented on Bob Gray's website through the link, http://www.rwgrayprojects.com/Lynn/Lynn01.html, and described as a whole in [2]. From the geometry, topology and the dynamics of the Mereon Matrix, a template information model was elicited through matching model characteristics with systems behaviour combined with in between small scale pilot applications. An early publication of the evolving model is provided in [8]. Since the year 1995, numerous highly esteemed scientists of all professions contributed to an initial exploration of what Mereon really is, cf. the series of Sequoia Meetings that are video documented, while only a small international team of researchers has contributed in a longer time span. This is seen as evidence of a scientific study, while also knowing that the research team is only at the beginning of its endeavour of exploring the Mereon Matrix and documenting what it is and what it may be successfully applied for. At present, there is a catch-22 situation, where colleagues repeatedly ask for peer-reviewed evidence, while the exploration is only in its beginning and the team is struggling to get funds for practical applications. Nevertheless, it is strongly believed that the present results are solidly founded in scientific practice.

### *1.2 An Essential Question*

While reading the review of Koonin, [9], it suddenly dawned on the present author that there might be a discrepancy in the meaning of the concept of 'evolution'. The trigger point was the text, "Analysis of numerous sequenced genomes vindicated Ohno's vision of gene duplication as a major evolutionary mechanism (*ref*) …" ([9] *page 1021*; the underlined represent the point in question). This is slightly deviating from the perception by Lynch in [10], expressing that there are four forces in evolution (natural selection, mutations, recombination and genetic drift), where the latter three "are non-adaptive in the sense that they are not a function of the fitness properties of individuals" ([10] *page 8597*).

The essential question derived is "what is biological evolution? ".

There is no doubt that the concept of evolution is an integrated part of the system of biological organisms, but does the concept of evolution include the mechanisms providing the variation based on which evolution exerts its effect? These includes mutations of all kind, gene drift and shuffling, duplication, and more, originating in meiosis, mitosis and some metabolic processes in general when failing, in particular everything related to the gene expression machinery. If the answer is yes, then evolution includes almost the entire molecular genetics. Therefore, a definition is needed.

Three definitions were found in dictionaries and Wikipedia, respectively:

- **"Evolution** is change in the heritable traits of biological populations over successive generations" Wikipedia (accessed 9th June 2016)
- **"evolution … 1.** *Biology*. a gradual change in characteristics of a population of animals or plants over successive generations: accounts for the origin of existing species from ancestors unlike them." ([5])
- **"evolution 1** *(in biology)* the process of cumulative change occurring in the form and mode of existence of a population of organisms in the course of successive generations related by descent." ([11])

Each of these definitions embraces the perception covered in this book, or at least they do not exclude the laws of physics (Quantum Mechanics, thermodynamics) that all processes have to obey, and thereby the analysis of pre-life evolutionary mechanisms remain open as an option within the present Unifying Theory of Evolution. However, none of the definitions are sufficiently detailed to clarify the above question. Therefore, the author has to decide for herself what she means with 'evolution', and this is:

> "Evolution is the combined set of principles and mechanisms that permanently and heritably implements the cumulative change occurring in the form and mode of existence of a population of organisms in the course of successive generations related by descent".

That is, for instance the variation (e.g. mutations) that is generated by meiosis and mitosis and errors in the gene expression system are NOT included in the system of evolution as such. Forces in evolution (mutations, recombination, and genetic drift) are perceived as fuel for the evolution – that is, generating possibilities or potential for evolution. Thus, they are factors in evolution but the processes producing them are not part of the evolution as such, and hence these are not 'mechanisms in evolution', while 'natural selection' is still perceived as an evolutionary mechanism.

Then, "What is 'natural selection'?"

The definition in [11] nicely expresses the present author's perception of the concept of natural selection:

> "**natural selection** the principle that the best competitors in any given population of organisms have the best chance of breeding success and thus of transmitting their characteristics to subsequent generations.".

The essential in this definition is that 'selection' is not an active process of choosing between alternatives, but a principle (see the formal definition of 'principle' in Section 2.2.3) that may have nuances to how it is interpreted in various contexts.

Further, the notion of 'chance' (in the definition) associates with a probabilistic principle.

Moreover, even if Darwinian theories have a particular perception of natural selection, the present author prefers to interpret the notion of 'natural' in the general meaning: 'natural' means "existing in, or produced by nature" ([11]), and in this perception, it does not enforce any particular type of imprint or prescription on a specific principle for the selection. Therefore, the nuances in the interpretation of selection are concerned with various factors contributing to the fitness of an organism/organisation within a context, for instance, purifying selection, relaxed selection, neutral selection and positive selection, where the adjectives are more or less self-explanatory.

Also, given the formulation in this definition, selection operates at an organismal and/or a population level, while the causal evolutionary mechanisms operate beneath this and at all levels: molecular biochemistry and genetics, physiology, morphology, etc., and even the laws of physics.

## *1.3 Delimitation*

Given the huge number of papers and theories, this study can only bring the essence of the various theories on mechanisms behind evolution. It is NOT the purpose to provide systematic reviews of the (sub-)topic(s) that are included in the model, not even a mini-review – only outlines of the recent literature, and only to achieve sufficient information for delineating each topic to a degree that makes the validity of the overall framework comprehensible and plausible. Note that this is intentional. The reason is that the certainty of accuracy in the final model does NOT come with completeness in coverage of every subtopic, but with a balanced coverage of the literature and a homogeneous level of detail.

Excluded is sexual selection as an independent factor of evolution, primarily to make the model simpler and hence more comprehensible.

While the modelling process is iterative and incremental and in principle involves all steps at each iteration, the individual steps will not be visible to the reader within the final model.

# 2 Methods and Basic Terminology

**Abstract.** This chapter is of the traditional type that describes the foundation for accomplishing a study/work, – that is, description of the theories, approach, material, and alike, including basic terminology and premises/conditions for the work and its derived conclusions.

**Keywords.** Methods, methodology, Mereon Matrix, template information model, perspectives

However, while the modelling work is founded on the Mereon Matrix, this need not be of major interest for the readers interested in evolution. Therefore, some details of this material is referred to Part 4 while the present part of the study report is keep fairly free from references to the modelling approach other than in general terms.

## *2.1 Outline of the Template Model*

The Mereon Matrix has provided us with a template for the modelling process, described in [2]. It consists among others of 7 interrelated functions:

A) The structural components coming into play – that is, readying the very foundation for evolution.
B) The dynamics in terms of the following three activities:
   a) Communicative interactions;
   b) Stabilisation (i.e. related to efficiency[3]);
   c) Internal regulation/prioritization (i.e. related to effectiveness);
C) Patterning/differentiation and fidelity (orchestration).
D) Evolvability of evolution itself.
E) Integration within the external environment.

**Figure 1** illustrates the principle of fixed succession of the 7 functions – that is, it is analogous to the functionality of Krebs' cycle – that is, cycling sequentially through the series of processes, however, in a spiral fashion since the input resources are incrementally modified over time by the sequence of invariant functions.

More detail on the Mereon Matrix may be found in Part 4, to which is referred.

Since the template has a fractal nature, each micro-function of a given function – that itself may be a micro-function of a higher level function – covers the exact same topic(s) as its macro-function but at an elaborate and dedicated degree of detail.

---

[3] 'Efficiency' is related to a measure of the capability of doing the things right, while 'effectiveness' is related to the capability of bringing about the result intended – i.e., doing the right things, under real circumstances; and 'efficacy' addresses the performance under ideal circumstances.

Additionally, the Mereon Matrix shows us three key concepts: Unity (in diversity), Perspective and Paradox, of which the concept of 'perspectives' is outlined in a dedicated chapter (Chapter 3).

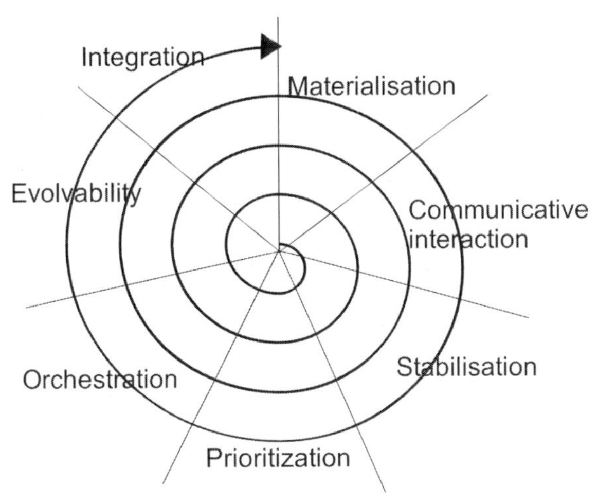


**Figure 1:** A spiral diagram illustrating the incremental nature of any system (incl. biological evolution) showing the sequential order of the functions traversed repeatedly, according to the Mereon Matrix.

## *2.2 Modelling Methodology*

The information modelling methodology is similar to methodologies for systems modelling in general. The difference merely constitutes the instrumental template information model and the properties of this. More detail on the modelling methodology may be found in Part 4, to which the interested reader is referred, but which may not be a particular interest to people from the application domain of biological evolution.

### *2.2.1 Step 1: Defining the 3rd Level Mereonic Function*

This step comprises a formulation of each sub-function of the 3rd level of the Mereon Matrix template model. It is visible only from Chapter 16, and may not be of any particular interest to a reader from the genomic application domain.

### *2.2.2 Step 2: Getting an Overview of the Knowledge Domain*

Getting a solid overview of the knowledge domain comes with a literature search (including textbooks), identifying the concepts and relating them to each other. In the present case, the entry point for acquiring the necessary overview was the insight and the literature acquired during our pilot study.

### *2.2.3 Step 3: Matching the Template Model with the Application Domain's Functionality*

We have two descriptions at our disposal: 1) the original function descriptions of the Mereon Matrix's macro-function and its first micro-level, see [2] (*pages 3-21*); and 2) the corresponding, interpreted description of the ATCG macro- and micro-functions in [1]. The latter, our pilot application, was concerned with the system of molecular genetics of living beings with a focus specifically on humans; we called this model an 'Applied Theory accounting for human moleCular Genetics' (ATCG), as it is a theoretical exercise "applied for a real case". Within this context, the present study constitutes the elaboration of one particular aspect (evolution) in terms of the model's next fractal level.

The purpose of this step is to identify the content – that is, to identify the headings of what to be included where. This naturally goes via an identification of Principles and Mechanisms within the functionality.

**Figure 2** illustrates in a diagrammatic fashion how each function operates. The functions in themselves are invariant and so are the principles and mechanisms, while the variant factors comprise the input resources (the potential) and the output resources (the emergent property of a function). Since each emergent property serve as input resources (renewed potential) in the cyclical iteration through all processes, this results in a system with resources of gradually increasing complexity, expanding the solution space. In a biological context, the system in focus comprises nature with various populations of species phenotypes.

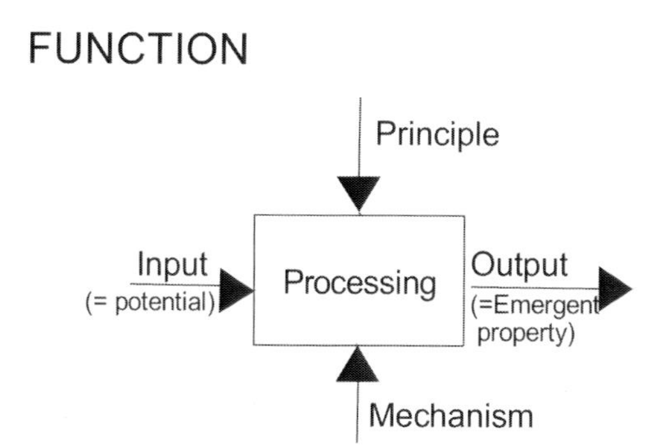


**Figure 2**: The diagramming type is slightly modified activity diagrams (based on IDEF0): boxes comprise activities / processes (i.e. 'functions' in a Mereon Matrix context); arrows into a box from left are input (resources serving as potential); arrows out to the right from a box are output (i.e. 'emergent property', modified resources); arrows downward onto a box comprise the guiding rules / competence for control of decision-making during the processing; and arrows upwards to a box are the acting mechanism / actors operating within the function.

The '**Principle**' corresponds to the essence of the function: "…**7**. a rule or law concerning a natural phenomenon or the behaviour of a system…" ([5]).

The '**Mechanism**' comprises the implementation approach for achieving the Principle: Boogerd and co-workers express the purpose of spelling out a mechanism as "… how some phenomena of interest–some reliably generated behaviour of the system–is generated by reference to how a number of components interact." ([12] *page 727*). The authors bring two definitions from the literature: "Mechanisms are entities and activities organized such that they are productive of regular changes from start or set-up to finish or termination conditions…" and "…a structure performing a function in virtue of its component parts, component operations, and their organization. The orchestrated functioning of the mechanism is responsible for one or more phenomena…" ([12] *page 727*).

### 2.2.4 Step 4: Filling Details for the Micro-Micro-Level Functions

While Step 3 of the modelling process identifies the 'headings', the present step fills the actual contents under each heading. Thus, it includes for each sub-function a description of the application domain's knowledge, formulated within the domain's terminology.

It is at this step that the perspectives (see Chapter 3) may become valuable in the modelling process, provided that one has sufficient details in the application domain.

### 2.2.5 Step 5: Defining the Emergent Properties

The 'Emergent Property' comprises the outcome of a function's operation. This step is accomplished iteratively and incrementally with the previous steps 2-4 and the following steps.

The important is that every (micro-) function elaborates on that which enables subsequent functionality, since the emergent property constitutes the input resources for the subsequent function. One example is that histone modifications enables coding for the alternative splicing: histone modifications constitute the materialisation functionality; coding constitutes a functionality of the communicative interaction; and alternative splicing belong with the regulatory (also including prioritization) functionality.

### 2.2.6 Step 6: Identifying Holes in the Model

Are there wholes – i.e. missing or insubstantial pieces of information – in the model? This could for instance have the following three causes of origin: 1) incomplete domain knowledge; 2) incomplete literature search; and/or 3) a flaw in the template information model. The purpose of this step is to identify and subsequently – if feasible – fill potential holes or otherwise remedy the cause to the problem identified.

The first two candidate risks are discussed in the Part 3, Discussion, while analysis of the last candidate cause is referred in its entirety to Section 18.1.

### 2.2.7 Step 7: Evaluation of the Model

This step comprises validation of the final model as coherent and convincing. Modelling based on the Mereon Matrix is judged to be successfully accomplished under the following two conditions:

1) That there is an adequate match at macro and micro levels between data and information of the knowledge domain and properties of the Mereon Matrix; for instance, a) is the progression from one micro-function to the next smooth, continuous and coherent; b) are there chunks of information in the knowledge domain that has not found their place in the model? c) … or the other way around, are there significant holes in the model? If so, identify the reason behind!
2) When Principles of the template information model are sequentially addressed at both macro and micro levels and at a sufficient level of detail

This validation is addressed in much more detail in Chapter 12.

## 2.3 *Quality Management of Input Sources*

Since no one is and cannot be domain expert in each and every niche dealt with at the continued detailed modelling of molecular genetics, the quality management of domain knowledge used as an input source of information had to be handled explicitly:

A) Literature search: The reference list from [1] served as a fairly up-to-date entry point. The literature search was iterative and incremental, using in particular the PubMed literature database. The search process iterated steadily between a) topic-oriented search (with synonyms), b) search of key authors, c) exploiting PubMed's concept of 'related papers' and 'topic reviews', d) tracing reference lists for significant papers on the topic in question, as well as e) numerous journal mailing lists that have been followed for years and hence have led to a growing personal knowledge base. The experience from [1] was that the literature of the domain of molecular genetics was extensive for almost every aspect dealt with, although more extensive for the hard-core biochemical aspects of human molecular genetics than the more 'soft' aspects also appearing within the domain of evolution.

B) Through consensus seeking: It is not the aim of this study to be exhaustive, but comprehensive, while appraising the scope and strength of each candidate paper. One of the lessons from [1] was that the speed of progress in the domain of human molecular genetics is huge, and therefore the one review follows or competes with another. Consequently, consensus seeking is made feasible by taking advantage of reviews. Since these reviews are neither meta-analyses in the Cochrane sense nor systematic reviews, there are no explicit inclusion and exclusion criteria related to for instance methods or biases or the like, neither in the literature nor in the search for information. To the extent feasible, a consensus will be deduced by ordinary scientific principles, while considering the fidelity of each paper.

C) Age of references: Again, based on the experiences from performing the modelling work included in the Mereon Matrix book, the focus will be on recent literature to secure most recent insight, with the exception of some classical papers (Darwin is one obvious example) and some key authors.

D) Quality of papers: This criterion is in general more subjective:

   a) The handling of references as evidence (cf. [13]). Of particular importance in the present context is the appropriate transmission of uncertainty of conclusions within papers, taking their methods, potential bias (e.g. in terms of inclusivity), etc., into consideration in the resulting citations. Citations are extensively used to secure accuracy of statements while attending to their context, but also to give full credit to such sources of evidence;

   b) Method bias was observed in particular through the harsh debates on some topics, and had to be handled explicitly;

   c) Most reference lists for papers within this knowledge domain are extensive, but in a couple of cases the age distribution of references within a given reference list was outdated compared to the publication date of the reference itself, and hence, this issue had to be handled explicitly to minimize holes in the model;

d) Observing the papers' self-assessment and their discussion of weaknesses and strengths of methods and conclusions, as well as assessment of assumptions' validity, as for instance is common in statistics and probability theory;

e) A final criterion is comprehensibility: a couple of papers were discarded because of repeated ambiguity of statements and lack of definition of key concepts. No articles were disregarded on the basis of only linguistic problems (loose speaking) such as in the following example: "Kin selection theory is a kind of causal analysis.".

### 2.4 *The Concept of Perspectives*

Our understanding of the concept of 'perspective' originates from the early days within the systems development domain where it was defined as "assumptions on the nature of the working processes and how people interact in an organizational setting" ([14]), see also the extended discussion in [15] from which this presentation is an excerpt. Implicit or non-conscious models of understanding behind the principles form the basis for our methods and management principles and thus for how we deal with things in a project/study. The concept of 'perspective' stands for hidden aspects and assumptions deeply buried in the design and application of methods. In a generalized version, 'perspective' is the implicit assumptions of (cause-effect relations within) the object of study. So, 'perspective' is synonymous with that aggregation of (conscious or subconscious, epistemological and ontological) assumptions of causal relationships, e.g. cause/effect relationships in a given object, in combination with imprinted attitudes guiding our decision making. This should be seen in the context of our – often subconscious – culturally conditioned way of perceiving a situation and interpreting observations. Few method designers are aware that our cultural background (professional, religious and national) alone maintains a series of tacit assumptions affecting our way of acting and perceiving things; see for instance ([16], [17], [18], [19]).

Caused by the tacit nature, some perspectives may contain pitfalls, where the perspective of a method conflicts with the actual purpose, which the method is intended to be used for. This is why this concept is so important in a modelling context.

The Mereon Matrix has three symmetry axes along which the structural complexity appears different (see Chapter 6 in [2]), even if it is the same object looked at, while looking at it from those three different 'angles', like a hologram. In an information model context, this corresponds to three different perspectives.

An example in [2] (Chapter 2) from modelling the Mereon Matrix shows that the above impact of 'perspective' is the case not only for somewhat abstract decision-making, but also in a concrete physical context of observation. Actually, one only needs to think of paintings to get that the painter's angle of observing an object impacts the picture, although the resulting model on the canvas is also coloured by the state of the artist's mind and his/her mastered techniques. The conclusion that came out of the observation from modelling the Mereon Matrix was that the different 'angles' of observation of a single object are equally valid and coexist. Therefore, the Mereon Matrix's template information model suggests that a pluralistic approach likely is the accurate one.

# Part 2: The Unifying Theory of Evolution

The purpose of this part is to provide a framework that encompasses a representative set of theories on evolution achieved since Darwin's publication of his theory more than 150 years ago in order to illustrate how these theories plug into a single shared framework and together create a coherent wholeness, the Unifying Theory of Evolution. It is not the purpose here to provide a review of the individual theories of evolution within the literature or to reconcile them, only to describe them to an extend that justifies their role within the wholeness.

# 3 Perspectives in Evolution

**Abstract.** The various schools of evolutionary theories are evident in the literature on biological evolution. By means of an existing template information model, the author succeeded in unifying all evolutionary theories into one shared framework consisting of three perspectives for each of seven functions that operates in a sequential manner. This chapter presents the overall Unifying Theory of Evolution, while subsequent chapters put detail on each of the functions and perspectives.

**Keywords.** Mereon Matrix, template information model, perspectives, evolution, Standard Evolutionary Theory, Extended Evolutionary Synthesis, Inevitable Evolution Theory, Unifying Theory, natural selection, survival of the fittest

The essence of Darwin's evolutionary theory may be summarised in terms of the following excerpt "As many more individuals of each species are born than can possibly survive; and as, consequently, there is a frequently recurring struggle for existence, it follows that any being, if it differ however slightly from its population of fellow beings in any manner profitable to itself, under the complex and sometimes varying conditions of life, will have a better chance of surviving, and thus be naturally selected. From the strong principle of inheritance, any selected variety will tend to propagate its new and modified form." ([20] *pages 4-5*).

"A widely accepted definition of evolution is change in the genetic composition of populations, ..." says ([3] *page 5*), while explaining the traditional view on evolution. Such definition unfortunately is operational by nature in that the enumerating part of the definition is expressed in terms of measures or measurable characteristics of the subject defined – that is, the definition *per se* excludes other perspectives and new understandings and thus induces a circular inference in the derived science, with the unfortunate side effect of debilitation. Therefore, a better – and functional – definition of evolution was searched, found among others in [11] and subsequently elaborated; see Section 1.2: "Evolution is the combined set of principles and mechanisms that permanently and heritably implements the cumulative change occurring in the form and mode of existence of a population of organisms in the course of successive generations related by descent". This definition leaves the opportunity for exploring various causal explanations to Darwin's evolutionary theory and theories gained in the period following him.

The deep controversies among schools of evolutionary theories quickly became prominent to the author. The viewpoints of each of the opposing parties all seemed justifiable, as argued in their respective contributions to the domain literature. An elucidation was needed. For the reader, the below will explain that the author of the present paper is neither pro nor con either of these competitive theories but rather have a pluralistic (dialectical) view on evolutionary theories, where 'opposites' are in a mutually constituting relationship, opposites are united and interpenetrate each other. Laland

and co-workers say "a plurality of perspectives in science is healthy, as it encourages consideration of a greater diversity of hypotheses, and instigates empirical research, including the investigation of new phenomena." ([3] *page 10*). That is, among the perspectives there is not a single winner that has priority over the others; they co-exist in a constructive relationship.

The 'controversy' that ignited the idea of incorporating the characteristic of three perspectives from the template information model into the modelling work is noticeable from Laland and co-workers' commentary in Nature, [4]: Devotees of the Extended Evolutionary Synthesis express that they can contribute to the understanding of evolution by means of their theory, while the devotees of the Standard Evolutionary Theory claim that they can fully and appropriately explain the entirety by means of their models; this includes taking advantage of the properties of plasticity within their perspective on evolution as 'smoothening' the fitness landscape as phrased by Frank in [21]. Both parties are right – that shows the paradox. They are both right with respect to each their observations and subsequent explanation models (provided that objective scientific criteria are met in the various scientific studies). What is wrong is the attitude that these theories cannot coexist as evolutionary factors in their own right, and this is what the perspectives in the Mereon Matrix shows. From a Mereon Matrix perspective, the various theories are not (necessarily) mutually independent, nor mutually exclusive, but may be cooperative principles and mechanisms.

The paradox in perspectives is illustrated within [2], and the paradox vanishes the minute one sees and understands its causal origin. An analogy that might ease understanding the coexistence of different perspectives is that of the wave and particle properties of light; one cannot observe both at the same time, but they are nevertheless equally valid and coexist at all times. That is, one observes the same object/system from two different 'angles'. This paradox in scientific discovery showed the present author a new perspective on the application domain of evolution and its theories and controversies. In the context of evolutionary theories, it means that the contrasting theories are not necessarily mutually independent, nor mutually exclusive but cooperative principles and mechanisms.

The Mereon Matrix shows that there are three perspectives. Immediately noticeable from Laland and co-workers' commentary in Nature, [4], are the Standard Evolutionary Theory and the Extended Evolutionary Synthesis. Therefore, what is the third perspective? It might be Witting's 'Inevitable Evolution' in [22]; see Section 4.1. An alternative could be the 'Constructive neutral evolution' described among others by Stoltzfus in [23]. The underlying theory and assumptions of Witting's model and the constructive neutral evolution model are both sufficiently different from those of the other two perspectives, but neither can cover the entire perspective on its own from Function 1 to Function 7. Further, since the other two perspectives are based on genes, it has to be a theory that will enable us to find an explanation to pre-life evolution (discussed in Chapter 11) and which still fits into the model. Therefore, the 'inevitable evolution' is the preferred candidate. Unfortunately, Witting's theory is yet so new and controversial that it has until now gained insufficient attention to be verified by other research groups, so the proof of this choice comes with the modelling – it will soon fail in the modelling process if it does not work as the third perspective. The modelling is continued with his theory as the third perspective.

Now is the time to summarise how evolutionary aspects are grouped according to this viewpoint of diverging perspectives.

A very first and simple abstraction of the three axes immediately came up and is elaborated in tabular form in Table 1:

A) According to the Standard Evolutionary Theory (abbreviated 'SET'), evolution is based entirely on changes in the genome (i.e. DNA), the contents of which is often denoted 'information', and changes in biological functionality is based on changes in DNA. Thus, this perspective could be called **'information-based'**.

B) Without attempting to define knowledge, 'knowledge' is perceived as information processed within a context. Epigenetics is the regulatory principle for applying the information represented within the genotype throughout the life-cycle of the biological system; therefore the term **'knowledge-based'** is suggested to capture the essence of the Extended Evolutionary Synthesis (abbreviated 'EES'); however, this should not be confused with the computer science term 'knowledge-based' systems for systems based on artificial intelligence. When looking at the rightmost column in Table 1, one sees that there is a 'learning' (adaptability) element in each part of their functionality.

C) Then to complete the metaphor, the 'Inevitable Evolution' theory (abbreviated 'IET') will be the **'data-based'** axis in evolution. Looking at the functionality one sees that the functionality actually is based on something comparable to the concept of 'data': energy, density, population data, and so on, - that is, basic physical measures as opposed to anything interpreted within a context. Therefore, in analogy, this also includes instinct-based actions /behaviour at the higher functions.

Another and perhaps better way of characterising the three perspectives is in terms of their key features: **constituent**, **acquired** and **adaptive** properties, for IET, SET and EES, respectively. These distinctions of the views on the perspectives have been useful in identifying the positioning of various evolutionary theories. A further note, this distinction between 'acquired' and 'adaptive' is not conflicting with the characteristics in [24] (and others) that SET is based on 'random' mutational changes and that evolution from the neo-Darwinian perspective is acquired through an adaptive process where the end result arises from subsequent selection while evolution within EES's perspective is 'directional' (non-random, intentional) – that is, the former type of adaptation is retro-active/reactive while the latter is pro-active in its adaptation. This is an important distinction.

Note that by this categorisation in the three perspectives there is no intention to diminish, disrespect or favour any of the three axes. They are complementary with each their right within the wholeness.

Also note that as they are merely different views on what is going on at the same time, they operate in parallel and may interact with each other. Further, the outcome of one of them at the execution of a given function may be applied (exploited) by the others in any subsequent function.

The above views on the three perspectives were used intensely in the modelling and gave valuable inspiration for structuring the information and filling holes, the result of which constitutes Table 1 in its present form. Note that there are deliberately more than one appearance of some of the concepts (or a variety thereof); they are analogous, yet different. In some cases, following the identification of an empty cell in the table, pieces of information were identified in the literature, while in other cases a dedicated supplementary literature search was performed and resulted in actual studies on the topic.

The multiple appearances of single aspects, like 'collective decision-making' that appear for both Inevitable Evolution and the Standard Evolutionary Theories, need a little explanation. For Inevitable Evolution Theory, the collective decision-making may be based on a system of connected stimuli and response in single celled organisms or the forces driving the seasonal migratory navigation of birds. For Standard Evolutionary Theory, an example might be the decision-making mechanism within a group of lions foraging for food. The decision-making patterns corresponding to the axis of Extended Evolutionary Synthesis will be of a more complex nature and is expected to have a stronger element of learning; therefore, the cultural aspects of decision-making are assigned under this axis. Further examples are found under the respective functions.

Implicit from the template information model veiled behind the table, it is now obvious that an abstraction of the emergent property of the three 'parallel' perspectives on a sub-function has to be identical, and their 'union' of actual output from the sub-function's operation constitutes the input for the subsequent function's three parallel perspectives.

**Table 1:** Overview of the functionality assigned to the three paradigmatic evolutionary theories. See more explanatory details in the main text. Even if the text in the headline says "Neo-xx philosophy", this should not be interpreted literally, as the Unifying Theory of Evolution goes beyond and extends such constraining interpretation.

| Function | Inevitable Evolution Theory (IET) | Standard Evolutionary Theory (SET) | Extended Evolutionary Synthesis (EES) |
|---|---|---|---|
| | **Philosophy**: A basic driving force behind evolution is 'Energetic state'[4], i.e. Laws of physics, like e.g. Quantum Mechanics and thermo-dynamics | **Neo-Darwinian philosophy**: The basic driving force behind evolution is all kinds of genotype changes. | **Neo-Lamarckian philosophy**: Basic driving forces behind evolution is a) a force adapting animals to their local environments and differentiating them from each other, and b) a force driving animals from simple to complex forms. |
| 1 | Maturation of resources, also depending on:<br>• Self-organised criticality<br>• Punctuated equilibria | Maturation of the phenotype from the genotype, including:<br>• Self-organised criticality<br>• Punctuated equilibria | Maturation of the phenotype from the genotype, including:<br>• Plasticity<br>• Modularity and supergenes<br>• Genetic stabilisation<br>• Genetic accommodation |

[4] Witting in [22] does not define precisely what he means by the 'energetic state' of an organism. The best guess is for instance the laws of thermodynamics that anyway are applicable for all (bio-) chemical processes.

| | | | |
|---|---|---|---|
| 2 | Inter-individual interactive behaviour/dynamics, regulated through density-dependent (competitive) interaction:<br>• Quorum sensing<br>• Electrical signalling<br>• Frequency-dependent competitive interaction | Inter-individual interactive behaviour/dynamics, regulated through individual interactive behaviours, kin selection, including:<br>• Altruism<br>• Greenbeards<br>• Cooperation & Reciprocity<br>• Aposematism<br>• Prokaryotic adaptive immunity system | Inter-individual interactive behaviour/dynamics, - adaptive processes regulated through mutual interactions and communicative interaction with the environment:<br>• Plasticity<br>  o Epigenetic foundation for behavioural patterns<br>• Prokaryotic adaptive immunity system |
| 3 | Meta-population dynamics:<br>• Flock dynamics | Meta-population dynamics:<br>• Group dynamics, multi-level selection<br>• Group behaviours<br>  o Cheating and deception in the context of group dynamics and cooperation | Meta-population dynamics, adaptive processes:<br>• Acquired team behaviours, social plasticity |
| 4 | Constitutive (instinctual /reflex-based) collective decision making – without learning:<br>• Collective movement patterns, such as:<br>  o Shoal/flock behaviour<br>  o Sheltering<br>  o Dispersal /migration | Collective cognition / acquired rule-based decision-making preferences:<br>• Collective action patterns, such as:<br>  o Dispersal /migration<br>  o Fission-fusion societies<br>  o Misc. decision-making aspects | Collective cognition, acquired value-based preferences in decision-making strategies:<br>• Culture, including<br>  o Religion<br>  o Institutions<br>  o Politics<br>• Informed (adaptive) dispersal / migration |
| 5 | Constituent properties in orchestration:<br>• Interspecies dynamics, including:<br>  o Symbiosis and parasitism<br>  o Interspecies cheating and deception<br>  o Interspecific killing<br>• Time & timing<br>• Fidelity | Acquired properties in orchestration:<br>• Speciation<br>  o Developmental symbiosis<br>• Specialisation<br>• Time & timing<br>• Fidelity | Adaptive properties in orchestration:<br>• Speciation<br>• Specialisation<br>  o Eusociality<br>  o Social patterning<br>• Time & timing<br>  o Life-history theory<br>• Fidelity |

| | | | |
|---|---|---|---|
| 6 | Constituent processes in evolvability:<br>• 'Inevitable evolution' as a result of stochastic variation | Acquired processes in evolvability:<br>• Evolution of 'evolvability' in a complex space of combinatorial opportunities<br>o Population genetics<br>o Horizontal gene transfer<br>o Selfish genes | Adaptive processes in evolvability:<br>• Homeorhesis<br>o Population genetics<br>o Constructive neutral evolution<br>o Horizontal gene transfer<br>o Selfish genes |
| 7 | Constituent integration processes:<br>• Niche construction<br>• 'Resource enhancement altruism' | Intentional integration processes:<br>• Niche construction<br>o Reciprocal niche construction | Adaptive integration processes:<br>• Niche construction<br>o Reciprocity<br>o Technology factors |

Note that the model is 'linear' when followed for the various perspectives individually (because the traversal from Function 1 to 7 is sequential), but transactional when viewing the perspectives horizontally for a given function. Also note that neither perspective has priority over any of the other two.

We have kept the literature's concept names rather than renaming them, even if this may appear odd in some places.

Be aware that these three axes were not visible at the higher levels of the model – that is, in [1] – given an insufficient level of details.

Witting writes "The proposed process does not deny a role for contingency in evolution by natural selection, it finds only that the traditional view is incomplete ..." ([22] *page 260*). So, he knows that the paradigmatic evolutionary theories are not mutually exclusive, but co-exist. Witting even says that "At a first glance it is probably fair to say that many patterns of large-scale evolution on Earth can be argued to be consistent with both contingent evolution by historical selection and inevitable evolution by deterministic selection." ([22] *page 262*). Thereby, he expresses a viewpoint shared with the Mereon Matrix, namely that different perspectives may each explain certain properties of a system; however, one perspective may be better at explaining some parts and another perspective may be better at explaining another part of the entirety.

Note that the functionalities listed in the table gradually zooms out from functions related to individuals, to group-internal aspects (a localized population), to aspects of a group as a part of the global setting; all of that while nature decides who to select as the fittest. Also note that different strategies face different evolutionary constraints (selection pressures).

When going back to the pre-life period, evolution must have been driven solely by the laws of physics and stochasticity, because there was no DNA (or RNA for that sake) that served as carrier of information in the evolutionary sense. This does not imply that the Standard Evolutionary Theory was absent or non-existent at that time, but it had no input resources to operate on and hence left no trace of its existence. At some point in time, RNA existed as the only type of organismal genetic material, so also this period should be taken into account when evolutionary mechanisms are discussed. Moreover, there must have been a time in evolution where plasticity didn't exist and hence had no role in evolution, simply because it had not yet been generated at evolution. So, one cannot justify rejecting plasticity today as an evolutionary theory in its own right. The three perspectives of functions co-exist throughout time, although one or more of them may presently (or at some time in the past) take the extreme manifestation of 'nil' for a given function for the presently known functionalities in evolution. An alternative (or supplementary) explanation to empty cells in the Unifying Theory of Evolution may be that there are theories and mechanisms that exists at present but which have not yet been revealed to the evolutionary scientists.

Thus, given the assumption that the Mereon Matrix template information model is valid, in case the author is unable to fill in all the blanks this does not necessarily mean that a given functionality does not have a content, only that the author personally hasn't identified it yet within the literature, or that researchers have not yet reported it within the generally available literature that is indexed in the literature databases.

Since all three perspectives are present at all times, each species can have all three kinds of decision-making. Here, an example may be valuable: Individuals of the human species[5] may in different contexts show instinctual decision-making (e.g. some defence mechanisms, and at caring for progeny), rule-based decision-making (e.g. at executing craftsmanship) and value-based decision-making strategies (e.g. in culture). However, since the various species are different with respect to their bodily features, one should not expect all of the species to exhibit all three kinds. Moreover, for instance the development of cognitive abilities will make a highly significant difference in some of the functions, like Function 4, which is concerned with decision making.

Finally, note that resources within the system of evolution are shared among the three perspectives all along the processing/progress at traversal through the functions. Consequently, as an example, a function that operates according to the Standard Evolutionary Theory acts upon the emergent property from the previous functions, such as the outcome of plasticity.

Since the three axes of functions are independent, co-exist and cooperate, one may explain the performance of the system as follows: The axis able to best, fastest, and/or with least cost or risk (in some context) while processing the available resources is the 'winner' and consumes (most of) the resources, in a manner analogous to water running downhill where the landscape is lowest; and where there is a dead-end valley, the water will accumulate until higher than the lowest pass. Think fractal, the resources are processed according to aptness of the functionality available at that level of functions within the system.

---

[5] Mayr's definition of 'species' is "a system of panmictic populations that are genetically isolated from other such systems" ([24] *page 5*) – that is, they are able to interbreed freely to produce fertile offspring ([11]).

# 4 Materialisation: Readying for the Evolutionary Pressure, Survival of the Fitted

**Abstract.** Function 1, materialisation (see Table 2 and Table 3 in Chapter 14), is concerned with the system-internal readying of the system's resources for the evolutionary pressure – that is, establishing the phenotype. In this process, unfit resources are deselected, and at the same time the maturation process bring about the potential of the phenotype adapting to its environmental conditions. Thus, the principle of this function is 'survival of the fitted, fitting to survive', and the mechanism is 'natural selection of unfit phenotypes during the maturation of phenotype traits', explained for the three perspectives: Standard Evolutionary Theory, Inevitable Evolution Theory and Extended Evolutionary Synthesis.

**Keywords**: Natural selection, survival of the fittest, Standard Evolutionary Theory, Extended Evolutionary Synthesis, Inevitable Evolution Theory, developmental biology, ontogeny, epigenetics, adaptation, self-organised criticality, punctuated equilibria, plasticity, modularity, supergenes, genetic stabilisation, genetic accommodation, genetic assimilation, genetic compensation, transgenerational epigenetics, cryptic variation

Here, the study of Wang and co-workers ([25]) comes in, deselection of obvious fatalities – that is, the non-fit phenotypes are aborted and will not participate in the subsequent evolutionary selections. They report that the success rate of a pregnancy (in humans) is less than two thirds for all conceptions detected at Day 16 after conception. It seems justifiable to state that the most severe (lethal) genetic changes are naturally aborted first, meaning that the total rate of success may even be significantly smaller than the mentioned two-thirds during the early pregnancy. As some non-viable changes may survive the threshold of Day 16 detection of a pregnancy, the numbers tell us something about the remarkably high error rate. One could call this 'purifying selection' in its literal sense. Evolution is not the result of imprecision, but of natural experimentation. All in all, this indicates that Function 1, developmental biology, is a gate keeper; non-viable resources are dismissed and will not enter further into the system.

Following this, de-selection comes a continuous development to increase the fitness toward given external and internal conditions. Such development is laid out in previous generations; and in this process further adaptation to external conditions will be advantageous. Therefore, in addition to genomic changes plasticity and epigenetics are perceived as founding factors to bring in.

This leads to the formulation of the principle and mechanism for this function:

**Principle**: Survival of the fitted
**Mechanism**: Natural exclusion of unfit phenotypes during maturation of phenotype traits
**Emergent property**: Selected, matured phenotype

Mechanisms generating variation are not all mutually exclusive and may all have contributed to who/what we humans – as well as other species – are today. Combining the methodological opportunities for evolution will lead to an explosion of the solution space. The principle 'Survival of the fitted' will secure that phenotypes unfit for the environment will not survive and so are sorted out, while those with tolerated non-lethal changes will withstand within the population of beings and may become silent phenotype properties until they specifically turn out to be advantageous or the opposite.

Stochasticity – that is (in the present context), biological noise (molecular and/or phenotypic variance), in any biological process is a factor that cannot be ignored in relation to evolution, since it may increase the variation that is the basis for the process of selection, see for instance the review of [26] and dedicated papers, like [27], to find more on this topic.

Here, even if the emergent property is stated as 'matured phenotype', it may nevertheless be matured at many levels – just think of the state transitions from a new-born baby, over child, adulthood, adolescence, and till senescence, which are all in themselves stable states while a shift from one state to the next is characterised by a series of smooth changes in biology that are all available for evolutionary pressure, and so is the resulting phenotype at each stage.

| **Inevitable Evolution Theory (IET)** | **Standard Evolutionary Theory (SET)** | **Extended Evolutionary Synthesis (EES)** |
|---|---|---|
| Maturation of resources, also depending on:<br>• Self-organised criticality<br>• Punctuated equilibria | Maturation of the phenotype from the genotype, including:<br>• Self-organised criticality<br>• Punctuated equilibria | Maturation of the phenotype from the genotype, including:<br>• Plasticity<br>• Modularity and supergenes<br>• Genetic stabilisation<br>• Genetic accommodation |

### *4.1 Physical Foundation behind Evolution, 'Inevitable Evolution Theory'*

The basic mechanism for evolution according to Witting in [22] selection is operating on energetic states, which the present author interprets broadly as the laws of nature and physics, such as the laws of thermodynamics and quantum mechanics.

The metaphor of a landscape and the example of prions may illustrate what is meant by energetic state: A prion is a subgroup of proteins with the specific property that it can exist in various structural conformations of which one is pathological and has a self-catalytic ability, meaning that it can promote conformational change of other samples of the same protein to acquire the same pathological conformation. Each of the multiple conformations corresponds to a specific energetic state, and thus one may view these as appearing within a landscape of energy states. Since the prion state is irreversible and easily accessible under certain conditions, its energy state – in the mountain landscape metaphor – would correspond to a deep valley hidden behind a ridge of mountains, from where it is impossible to escape, while other conformational states may be switched from one to another.

Witting's idea regarding a basic driving force behind evolution is the 'energetic state', about which he says "A new formulation of thermodynamics in biological evolution was suggested by Pross (*refs*). He noted that the replication reaction is an extreme expression of kinetic control where the thermodynamic requirements play a supporting, rather than a directing, role." ([22] *pages 283-4*). Thus, his statement of 'energetic state' is perceived as the laws of nature (e.g. thermodynamics and quantum mechanics), and that these are elements behind the processes involved in evolution: For molecular systems, according to thermodynamics, the landscape of energetic maxima and minima of molecular chemical and physical processes determines a natural sequence of transitions (whether or not catalysed) that developments are forced to follow, and hence they define the path of selected transitions. Thus, such developments are not by-chance-changes and hence not random system developments but a change driven by natural forces.

Witting's idea of an evolutionary principle based on energetic state is not that far from the argument that all biochemical processes must obey the laws of nature, and that this therefore has to include all processes behind evolution; it will explain and make incorporation of the pre-life evolution into the theory of evolution feasible (see Section 11 within the Discussion). So, it is likely that his suggested evolutionary theory may have a role in evolution, and in such case it belongs in this perspective, starting with theories related to maturation of resources.

In summary, the readying of resources (materialisation) specifically in the context of Inevitable Evolution Theory is concerned with that part of the development of the phenotype, which takes place according to the principles of the laws of physics. However, whether it is a major or minor player in evolution is another question that the author will abstain from discussing.

### *4.1.1 Self-Organised Criticality and Punctuated Equilibria*

Note that the mechanism behind the theory of self-organised criticality and punctuated equilibria is based entirely on the laws of physics – although taking 'advantage' of the outcome (accumulated mutations) from previous functions/traversals of functions, and this is why it is included both for the Inevitable Evolution Theory and for the Standard Evolutionary Theory in Section 4.2.1.

Bak and co-workers (in [28], [29]) discuss self-organised criticality and punctuated equilibria in relation to evolution. Punctuated equilibria here serves as an evolutionary theory suggesting that the evolution for a given species as a whole shifts between phases of stasis (where latent changes may accumulate) and phases of very rapid morphological change (denoted 'devils' staircase'), based on accumulated mutational changes. Punctuated equilibria take place when the local change in the fitness landscape (for the species as a whole) has developed to a point that requires adaptive activity ('avalanche').

The mechanism behind punctuated equilibria is the self-organised criticality; Bak has previously illustrated the principle with the metaphor of a sand pile on which new grains of sand are continuously and slowly poured and as a consequence cause 'avalanches', each time the system reaches a critical state (see a brief account in [29]). The sand grains (in the sand pile metaphor) may be compared to mutations and the system to an evolving organism; according to this model, the spontaneous transition arises as a result of accumulated simple local interactions / events, and leading to complexity.

There seem to be an emerging consensus that this is a general phenomenon in nature and hence also in biology and evolution; it has been investigated for forest fires, earth quakes, and biological systems like nervous systems in terms of cell cultures, brain slices, and anaesthetised rats. However, although the experimental evidence of self-organised criticality as the fundamental mechanisms behind neural systems has been investigated for some years, it is still controversial; see for instance the review of [30]. Still, these authors conclude from their review that (for brain functions) the model of self-organised criticality as a mechanism of the brain is both feasible and plausible, and preferred over other alternatives. And, if it is a general principle, then it is definitely possible that we will also find it elsewhere in biology, which may include evolution.

In short, there is no conclusive evidence at present of punctuated equilibria and self-organised criticality as elements in an evolutionary theory as suggested by [28] and [29], but indications suggest that this is both feasible and plausible.

### *4.2 Genotype Change as Driver of Evolution, 'Standard Evolutionary Theory'*

The basic mechanism for evolution according to the Standard Evolutionary Theory is selection operating on mutation-based fitness characteristics of the phenotype, corresponding to an opportunistic trial-&-error with no a priori principles to determine directionality.

The essence of the Standard Evolutionary Theory is excellently summarised in Laland and co-workers' commentary in the journal Nature, [4], and in a follow-up by the same group in [3]. The essence is that the story, which the Standard Evolutionary Theory tells, is simple: "new variation arises through random genetic mutation; inheritance occurs through DNA; and natural selection is the sole cause of adaptation, the process by which organisms become well-suited to their environments." ([4] *page 162*). Thus, the conception behind the Standard Evolutionary Theory is that the mechanisms behind evolution are based entirely on accumulated mutations. The characterisation in [4] continues with "In this view, the complexity of biological development – the changes that occur as an organism grows and ages – are of secondary, even minor, importance."; – that is, according to the Standard Evolutionary Theory, the sole driver of evolution is changes occurring through DNA, and that the arguments of devotees of the Extended Evolutionary Synthesis (such as niche construction, plasticity, and epigenetic inheritance) are already well integrated in evolutionary biology and therefore does not need a separate theory.

Back to the topic of genotype changes as drivers of evolution: Bihlmeyer and co-workers in an impressive 17 cohort study collected worldwide ([31]) conclude that "This effect was consistent between European and African Ancestry cohorts, men and women, and major causes of death (cancer and cardiovascular disease), demonstrating the broad positive impact of genetic diversity on human survival." ([31] *abstract*), with an estimated mean decrease of a person's risk of death by 1.57 % per standard deviation of heterozygosity that an individual has above the mean value. So, they demonstrate that even a small genetic diversity is indeed an evolutionary advantage.

In this context, all mutations, relocations (translocations), gene drift, inversions, etc., serve as parts of the foundation for evolution; some arise during mitosis or meiosis, and others again at any other organismal (e.g. regulatory) processes that fail. So, the key point here is as said in [1] (*page 373*) and also cited above:

> "... Evolution is not the result of imprecision, but of natural experimentation; it is not intentional in a cognitive decision-making sense, but in how that when a renewed resource turns out to be supportive for the wholeness of the cell and its greater context (the body as an expressed phenotype) it is exploited to its max through selection."

(end of citation).

Inspired by Figure 1 in Innan & Kondrashov in [32], a possible mechanism for how the effect of punctuated equilibria and self-organised criticality come into play might be as follows: cryptic variation continues to accumulate in individuals who reproduce and thereby it will also accumulate in the population. At some point in time, a fate-determining event or condition happens that opens for the exploitation of parts of the potential within the accumulated cryptic variation, and which thereby becomes a visible part of the phenotype trait. Following this, a preservation phase starts to maintain and secure continued availability of that gene expression, while the new phenotype engages in the natural selection process.

In short, the readying of the resources at this perspective comprises maturation of the phenotype to a point making it applicable for the selection pressure, and in this respect also factors influencing this process in various ways.

### *4.2.1 Self-Organised Criticality and Punctuated Equilibria*

The mechanism behind the theory of self-organised criticality and punctuated equilibria is based on laws from physics, suggested by Bak and co-workers (in [28], [29]) as a factor in evolution. Since it is based on laws of physics, it is referred to the Inevitable Evolution Theory, but the present author find that it is also relevant for the Standard Evolutionary Theory, as it may provide a mechanism to explain sudden major evolutionary jumps based on the ability to accumulate mutations.

A little more information on this theory may be found in Section 4.1.1.

However, as also pointed out in Section 4.1.1, there is at present no conclusive evidence of punctuated equilibria and self-organised criticality as elements in an evolutionary theory, as suggested by [28] and [29], but indications suggest that this is both feasible and plausible.

## *4.3 Regulation of Gene Expression as Driver of Evolution, 'Extended Evolutionary Synthesis'*

As for the Standard Evolutionary Theory, the readying of resources here comprises maturation of the phenotype to a point that makes the phenotype ready for the selection pressure.

Beldade and co-workers introduce their article by the following simple and expressive statement that clearly contradicts the attitude among devotees of the Standard Evolutionary Theory: "It has become clear that the environment is more than just a filter of phenotypic variation during the transgenerational process of natural selection, as it also plays a key role in generating variation during organismal development." ([33] *page 1347*). After all, it is the phenotype that is the target of natural selection, not the genotype.

The basic mechanism for evolution according to the Extended Evolutionary Synthesis is selection operating on fitness characteristics arising as consequence of adaptive changes in the phenotype driven by organismal (e.g. regulatory) processes by epigenetics (including the effect of small regulatory RNAs). Thus, it constitutes a reactive adaptability (i.e. implicitly an a priori directionality) with directed selection.

Evolutionary developmental biology ('evo-devo') is closely connected to ontogeny (the biological development of the organism), and it is believed to bias and constrain evolutionary pathways; as referenced by Laland and co-workers in [34]. Development-minded evolutionists argue that development processes constitute significant but neglected evolutionary mechanisms in their own right, emphasising the roles of developmental plasticity in evolution, especially in the formation and prevention of novelty. However, Laland and co-workers also assert that "…it is difficult to reconcile conventional evolutionary thinking with the view that development must be regarded as an evolutionary process; a process that is not fully controlled by genes." ([34] *page 209*).

The discussion continues in the commentary in Nature, [4] and in [3], both of which explicitly present the two opposing views on mechanisms behind evolution. A citation from the Nature paper clearly expresses the Extended Evolutionary Synthesis point: "In our view, this 'gene-centric' focus fails to capture the full gamut of processes that direct evolution. Missing pieces include how physical development influences the generation of variation (developmental bias); how the environment directly shapes organisms' traits (plasticity); how organisms modify environments (niche construction); and how organisms transmit more than genes across generations (extra-genetic inheritance). For the Standard Evolutionary Theory, these phenomena are just outcomes of evolution. For the Extended Evolutionary Synthesis, they are also causes." ([4] *page 162*). The counterpart, the Standard Evolutionary Theory, similarly express their viewpoint "In essence, Standard Evolutionary Theory treats the environment as a 'background condition', which may trigger or modify selection, but is not itself part of the evolutionary process." ([4] *page 162*).

While the reality of epigenetic inheritance is now indisputable, its significance to the evolutionary process is less agreed upon by everyone. Epigenetics is a major factor in developmental biology and therefore also a factor determining the phenotype, which constitutes the target of natural selection, and which inevitably has an indirect or direct effect on other evolutionary factors. For example, more or less stable epigenetic inheritance that silences parts of the genome will influence the rate at which nucleotide substitutions occur (see e.g. [35]) and this way influence the mutational basis for the subsequent evolution.

Laland an co-workers in [3] summarise the essence of the divergent views between the Standard Evolutionary Theory and the Extended Evolutionary Synthesis in terms of six core assumptions: i) the pre-eminence of natural selection versus reciprocal causation; ii) genetic inheritance versus inclusive inheritance; iii) random genetic variation versus non-random phenotypic variation; iv) gradualism versus variable rates of change; v) gene-centred perspective versus organism-centred perspective; and vi) macro-evolutionary patterns determined by micro-evolutionary processes of selection, mutation and gene flow versus additional evolutionary processes like developmental bias and ecological bias (for more detail see also [4]).

The author's view is that developmental processes will certainly influence the evolution. One example is the epigenetic settings that regulate the developmental processes, differentiation, influence fertility, longevity and disease resistance, and more, and thereby throughout life prepares an individual for the evolutionary pressures; see also the next section. Influencing for instance fertility and disease resistance interferes with the very core of the concept of fitness. The developmental processes are the result of the ongoing conditions, internal and external to the system, while mutations, relocations (translocations) and more are continuously ongoing. Our metabolism constitutes one large network of connected processes – so the developmental processes will directly and indirectly affect the development from genotype to phenotype, and hence also the evolutionary processes.

Consequently, developmental processes and epigenetic should definitely be considered a factor in evolution.

We prefer to apply the notion of 'developmental biology' for readying system resources or their properties in terms of a state or conditions for these resources to serve as input for subsequent evolutionary processes. Moreover, this readying is perceived as an active evolutionary process and support a pluralistic perspective on evolution as this entire paper demonstrates.

In summary, the materialisation (also named 'readying') of resources for the evolution is concerned with the maturation of the phenotype under the given environmental conditions.

### *4.3.1 Epigenetics as an Evolutionary Factor*

Skinner defines epigenetics as "molecular processes around DNA that regulate genome activity independent of DNA sequence and are mitotically stable …" ([36] *page 1297*), or a similar dictionary version, "the study of factors that influence gene expression but do not alter genotype, …" ([11]). Skinner mentions a handful of mechanisms that implements epigenetics, such as DNA methylation (which probably should include also the other types of DNA modifications), histone modifications, chromatin structure and selected ncRNAs.

Methylation of DNA is an epigenetic modification of the genome that regulates crucial aspects of its function, for instance in relation to DNA repair, CGI-based regulation, and chromosome condensation during mitosis and meiosis, as well as organogenesis and morphogenesis. It represents a form of annotation mediating gene expression. Other kinds of epigenetic regulation are implemented through modifications of histone with one or more (out of a series of possibilities) of usually prosthetic groups on the amino acids in given positions on the histone variants in chromatin. From [37], it seems that there is interplay between histone modifications and the more stable DNA modifications (still reversible, but requiring several steps).

Epigenetic settings can be inherited over generations; see e.g. the book edited by Tollefsbol, [38], and also Skinner's paper, [36]. For instance, "A large number of environmental factors from nutrition to toxicants have been shown to induce the transgenerational inheritance of disease and phenotypic variation (*ref*)." ([36] *page 1298*), and this has been shown in living beings as diverse as plants, insects, fish, rodents, pigs and humans. Additionally, epigenetic processes are able to promote genetic mutations, if not even driving genetic change, and therefore, epigenetics can directly influence the phenotype traits as suggested by Lamarck; see also the discussion in Skinner, [36].

Valena and Moczek introduce their review with the short and strong statement "All developmental plasticity arises through epigenetic mechanisms." ([39] *abstract*) and they conclude with "Epigenetic mechanisms feature especially prominently in developmental plasticity and its evolutionary consequences." ([39] *page 11*), calling the process 'chaperoning action of epigenetic mechanisms'. Also other researchers refer to epigenetic mechanisms as the regulatory principle of plasticity; see for instance [40].

Infection with *Toxoplasma gondii* has several reported effects on behaviour, on man and more, such as effects on personality, change of fear, aggression and impulsivity; see for instance [41], [42], [43]: "…showing that, by hypomethylation of certain regulatory elements of key gene, *Toxoplasma* is able to reprogram the brain's genetic machinery." ([42] *page 5934*). Behaviour is involved in the mechanisms determining for instance altruism and cooperation (see Section 5.3.1), and therefore, behaviour is a variable in evolutionary processes, in particular in relation to the communicative interactions and to the effectiveness regulation.

Recently, the concept of transgenerational epigenetics has been acknowledged as a factor that directly affects fitness of phenotypes and hence evolution; see chapters in [38]. That is, the likelihood that external factors will impose an evolutionary process of selection on the foetus is definitely present.

The concept of 'meta-stable epialleles' (see [44]) denote the alleles (for mammals) in which variable expression is caused by epigenetic differences instead of genotypic heterogeneity, and which can exist in variable epigenetic states. Obviously, these meta-stable epialleles are a crucial factor in evolution, since they determine various traits in a phenotype, cf. the mechanism for implementing plasticity, for instance during the foetal development, but also later in life at acquired (non-infectious) diseases. The reviews [35] and [45] – and more in the same book – provide long lists with examples of epigenetic inheritance, transgenerational or through the maternal linage, and beyond.

Giuliani and co-workers in [46] suggest two different epigenetics-based mechanisms that could be involved in human evolution: a) selection-based effects, which are similar to the 'normal' gene-based selection; and b) detection-based effects that are accomplished through a detection process, which leads to the generation of and transmission of an epigenetic pattern apt to cope with the condition in question, and which would 'buy time' until a new potentially adaptive genetic mutation arises.

In summary, the evidence seems convincing that epigenetics is a factor in evolution in its own right.

#### *4.3.2 Plasticity*

"Phenotypic plasticity is the capacity of a single genotype to produce different phenotypes in response to varying environmental conditions" ([47] *page 459*); see also [3]. Thus, phenotypic plasticity refers to an individual's ability to respond to environmental changes by adjusting aspects of its phenotype (see among others the reviews [33], [39], [40], [48]), thereby emphasising the adaptive nature of phenotypes. Adaptive phenotypic plasticity enables organisms to cope with environmental variability to some extent, but also regulates the development or transition between the various alternative phenotypes depending on environmental clues. Thus, plasticity may be 1) an integrated part of the maturation of a phenotype, responding to the given conditions during the development, or 2) an active response to a change in the environment. The former role relates to the 'structural components coming into play', while the latter relates to 'communicative interactions' within the template information model (see Chapter 5); and consequently, these two aspects will be dealt with separately, here and for the following function, respectively.

Plasticity is inheritable and therefore itself subject to evolutionary mechanisms.

Aubin-Horth and Renn reference studies of actual investigation of affected genes and emphasise that "…there is often no single gene whose change in expression has the power to define a biological state." ([40] *page 3769*), and that modularity (see this concept later) has a key role, precisely because of the necessary coordination of specific gene expression. In a study, which the authors reference, 171 genes were involved in the plastic change. In another case, it was 15% of the genes studied, and in still another study 39% of the genome; this impressive latter example was concerned with the differential expression between the nurse and forager stage of honeybees.

Schlichting and Wund in [48] are more hesitant than [39] and [40]. They discuss briefly the literatures' suggested mechanisms underlying phenotypic plasticity characterised in terms of observations, for instance:

- Trade-off genes, respectively stress genes, for which the expression reflects the two extreme environmental conditions
- Genes whose patterns of expression change over time representing a core acclimatisation set shared among populations
- Genes with population-specific expression levels that do not change over time, with local adaptation
- Genes with shared time-dependent expression patterns and population-specific expression levels
- Genes whose plastic responses differed over time and among populations
- Changes in patterns of regulation of specific genes to provide adaptation to particular local environmental regimes

Schlichting and Wund in [48] conclude that the evidence of a causal link between epigenetic regulation and explanatory mechanisms behind plasticity is scarce. At the same time, the authors suggest that genes with plastic expression profiles are more likely to be targets of selection.

Another review from the same year, [49], also refers the mechanisms behind plasticity to the domain of epigenetics. Sharov divides the overall mechanisms into four groups, of which two are referred to the section on communicative interactions (i.e. Chapter 5), while the other two according to Sharov are:

1) Adjustment, simply the capacity to activate or repress a certain function; this is particularly relevant in the ontogeny;
2) Multitasking as the ability of sub-agents to handle multiple functions, for instance the ability of bacteria to quickly develop resistance to antibiotics because of flexible (multifunctional) enzymes – that is, an example of the adaptability to environmental changes; a similar example of developmental adjustment is the birch moth that changes colour when its habitat over time changes colour;

Valena and Moczek in [39] review the epigenetic mechanisms underlying developmental plasticity in their model species (*Onthophagus*):

- The regulation of gene expression
- Endocrine regulation; epigenetic regulation through DNA methylation. However, they say that this explicit regulation of plasticity is still poorly understood, yet they conclude that the combined data are consistent with the

hypothesis "that facultative methylation underlies adaptive, plastic responses to variation in nutritional environment." ([39] *page 8*)

- Conditional crosstalk between developmental pathways by means of a transcription factor

Beldade and co-workers in the invited review, [33], summarise a wealth of examples of plasticity and the mechanisms that implement such plasticity. Beyond the mechanisms already mentioned in the above, these authors further mention a) a "switch between alternative developmental trajectories that result in drastically different morphologies." ([33] page *1355*), but how it works seems little known at present; and b) modularity in developmental genetic networks, as decreased pleiotropy[6] between networks may facilitate the induction of different modules under different environmental conditions.

Ragsdale and co-workers study such a developmental switch that controls a morphological plasticity of the nematode adult stage and demonstrates that this developmental plasticity acts through a single enzyme, a sulfatase ([50]).

In short, the literature on plasticity convincingly indicates that plasticity is indeed an evolutionary theory in its own rights. There are many more and perhaps more convincing examples in the following chapters. We'll refrain from discussing whether plasticity has a major or minor role.

### *4.3.3 Genetic Stabilisation and Disruptive Selection*

'Stabilising selection' is defined as "It is environmental change that elicits a hidden portion of the reaction norm, with selection then favoring mutations that enhance responses to the environmental factor; finally, selection ultimately favors a stabilization of the reaction norm." ([48] *page 657*). Or: "**Stabilizing selection** Natural selection against individuals that deviate from an intermediate optimum; this process tends to stabilize the phenotype. By contrast, directional selection pushes it towards either extreme." ([51] *page* 505). The important in these definitions is selection favouring harmonisation of a population within the given context.

This type of selection decreases the variation of a phenotype within a population over time, simply through removal of the more extreme phenotypes within a population, and thereby making the population more homogeneous. Wikipedia (accessed 14th April 2015) mentions the good classic example of human birth weight, where small babies are disfavoured because of an increased rate of infections, and large babies are disfavoured because of a higher rate of birth complications; in both cases their mortality increases and the medium size is thus favoured.

Tautz cites a study showing "that enhancer elements are indeed subject to fast evolutionary changes but that stabilising selection can retain their functional conservation, by selecting for compensatory mutations." ([52] *page 576*).

'Disruptive selection' has the opposite effect of 'stabilising selection', namely favouring the extreme variants of a phenotype, probably because of heterozygote disadvantages or competitiveness among the average phenotypes. The effect of commercial harvesting of fish at high harvest pressure is one example shown by Landi and co-

---

[6] Pleiotropy is the "phenomenon in which a single gene is responsible for producing multiple, distinct, apparently unrelated phenotypic effects" ([11]).

workers in [53]. They explain that fishing policies can be selective for both size and maturity stage; the size specific selection is caused by mesh-size and gear regulation or from specific regulations, while maturity selectivity may arise when juveniles and adults are spatially separated during spawning.

### *4.3.4 Genetic Accommodation and Phenotypic Accommodation*

Genetic accommodation, assimilation and compensation comprise a set of mechanisms with which given phenotypic variants in an evolutionary trajectory are integrated genetically.

'Genetic accommodation' is defined as "a process by which phenotypic variants that are initially strictly environmentally induced are selected to become genetically determined (i.e. heritable)…" ([48] *page 657*); and leading to an increased frequency of that phenotype. Or, similarly: "… *genetic accommodation* … is a broad term referring to evolutionary mechanisms whereby selection acting on quantitative genetic variation moulds a novel phenotype, environmentally induced … into an adaptive phenotype …" ([33] *page 1353*). And further, "The concept of genetic accommodation describes trans-generational mechanisms of (quantitative) genetic change that can both fine-tune developmental plasticity or canalize development." ([33] *page 1353*).

"*Phenotypic accommodation* refers to the mutual and often functional adjustment of parts of an organism during development that typically does not involve genetic mutation [*ref*]" ([3] *pages 3-4)*, also promoting the emphasising that "phenotypic accommodation could promote *genetic accommodation* if environmentally induced phenotypes are subsequently stabilized and fine-tuned across generations by selection of standing genetic variation, previously cryptic genetic variation or newly arising mutations [*refs*]." ([3] *page 4*).

Note that "Evolution by genetic accommodation differs from the traditional view of the evolutionary process merely in that it begins with environmental perturbations, which through their effects on developmental processes alter the amount and nature of genetic variation visible to selection." ([54] *page 303*), so it is a proactive adaptation.

Then, as Moczek expresses it in [54]: "However, if *by chance* a certain environmental perturbation alters development in a way that it happens to produce an adaptive phenotype, and if *by chance* the same environmental perturbation results in the release of previously cryptic genetic variation, selection on which could stabilize the newly adaptive phenotype, then we have the principal ingredients in place for evolution by genetic accommodation to occur and to allow environmentally induced phenotypic variation to become heritable." ([54] *pages 301-2*); where he refers to 'cryptic variation' as "Individuals within a population can be genetically different from each other without leaving a signature of this difference in their phenotypes and reproductive success." ([54] *page 300*). That is, cryptic genetic variation is accumulated variation that is not phenotypically expressed under the given environmental or genetic circumstances [55], and hence not available for evolutionary processes.

"Genetic accommodation does not require new mutations to occur, but it might incorporate such mutations along with standing genetic variation, including variants that were formerly cryptic, neutral or rare in a population." ([55] *page 2706*). The mechanisms include feedback regulation, duplicate or redundant pathways, a balance between antagonistic processes and switch-like behaviour.

In the example of hypoxia tolerance in humans, the genetic accommodation (the fixing) is suggested to arise via mutations in genes regulating expression of plastic responses to hypoxia, perhaps through the constitutive production of the hypoxia-inducible factor signalling cascade. Genetic accommodation may be achieved for instance through plasticity or modification of the interrelationships between traits ([48]). The advantage of plasticity may be that it may enable survival in the changed environment and hence allow time for adaptation to such change by one or another kind of accommodation through regular selection mechanisms. Thereby the plasticity itself affects the evolutionary trajectory of the involved trait (and its correlated traits) with respect to form, expression, regulation, associated costs as well as the possibility of further integration with other traits or separation from these.

### *4.3.5 Genetic Assimilation and Compensation*

'Genetic assimilation' is defined as "a process by which a character state, produced initially by means of a plastic response, is sequentially fixed due to genetic modifications via selection that favors the loss of plasticity…" ([48] *page 657*). Or similarly: "*Genetic assimilation* describes an evolutionary process by which an environmentally induced phenotype becomes genetically fixed, so that the environmental cue is no longer necessary for the expression of that phenotype …" ([33] *page 1353*). Genetic assimilation is a sub-type of genetic accommodation.

Plasticity may be lost, for instance if its maintenance is costly (metabolically speaking), or because of relaxed selection when the alternative environments are not sufficiently confronted. Or, when a population is exposed to a novel yet relatively invariant environment then the new phenotype may become constitutive through genetic assimilation ([55]).

'Genetic compensation' is defined as "selection for similar phenotypes in different environments, achieved by divergence in underlying physiological plasticity." ([48] *page 657*), also a type of accommodation.

### *4.3.6 Modularity and Supergenes*

The hypothesis paper of Snell-Rood and co-workers, [56], discusses the role of modularity as a mechanism underlying phenotypic plasticity: they define modules "as semi-independent, dissociable units (e.g., genes, proteins, and traits), where interactions are more tightly correlated within modules than between modules." ([56] page *72*). This is immediately associated with the chromosome territories and the impact of transcription factories; see these concepts in [1]. However, the concept of 'supergenes' confirms and extends this association and the plausible mechanisms behind; see [57] and [58].

"Modularity in development permits entire networks or sub-networks to be induced by specific environmental conditions, or cues, through switch-like processes" ([56] *page 73*), so obviously modularity and plasticity may have a major impact on evolution – everything else equal – those able to adapt to the local conditions have better fitness. According to the authors, there are hundreds of examples of environment-specific gene expression, suggesting that modularity would be a common and good strategy for coping with environmental variation.

Once established, modularity favours the evolution of plasticity through reduction in pleiotropic constraints between alternative phenotypes ([56]), where 'pleiotropy' constitutes the correlated effects of a single gene on multiple phenotypic traits.

How modularity came about is addressed by Melo and Marroig in [59], who show 1) that directional selection is a prime candidate for the engine behind evolution of variational modularity whereas stabilising selection is critical for its maintenance, and 2) that drift is incapable of producing or maintaining variational modular structures for many generations.

'Supergenes' are defined as "A genetic architecture involving multiple linked functional genetic elements that allows switching between discrete, complex phenotypes maintained in a stable local polymorphism." ([58] *page 3*). Schwander and co-workers in [57] review a fairly large series of example cases (beyond the system of sex chromosomes), some of which may not (yet) fulfil Thompson and Jiggins' strict criteria for qualifying as a supergene: "...a biological systems needs to demonstrate clear evidence of a complex phenotype of multiple co-adapted elements, with a pattern of inheritance essentially identical to alternative alleles at a single locus, and maintained in a polymorphism in a single population." ([58] *page 4*). The two reviews [57] and [58], which by the way appeared within two weeks of each other, have slightly different angles on the topic, and both strongly argue in favour of the concept of supergenes; moreover, Thompson and Jiggins reference Hamilton for recognising that a supergene architecture was the likely mechanism for retaining the tight association between the signal and behaviour in greenbeards.

Schwander and co-workers in the review, [57], discuss several candidate mechanisms behind the emergence of supergenes, for instance related to: 1) clustering of loci, which for instance may arise as a consequence of gene duplication and/or translocation; 2) suppression of recombination (e.g. through location near centromeres; structural differences; epigenetic modifications; and chromosomal inversions – the latter seems key to supergene systems); and 3) maintenance of the supergene polymorphism (the fitness of the resulting phenotype that they regulate; frequency-dependent selection; and recessive lethal alleles in one of the supergene haplotypes).

Schwander and co-workers conclude their review with the statement "It will be of great interest to determine how frequently dimorphic phenotypes in a given species are produced purely by phenotypic plasticity, by polymorphic regulatory elements affecting the expression of several genes, or by supergenes whose sequence differences are directly responsible for the alternative phenotypes." ([57] *page R293*); indeed.

# 5 Communicative Interaction: Contributing Relational Aspects to Evolutionary Processes

**Abstract.** Function 2 is concerned with communicative interactions (see Table 2 and Table 3 in Chapter 14); in the present context of evolution that corresponds to the optimisation of fitness through the relationships between individual agents. The principle of this function is 'survival of the fittest', and the mechanism is 'communicative interactions guiding natural selection among kin', explained for the three perspectives: Standard Evolutionary Theory, Inevitable Evolution Theory and Extended Evolutionary Synthesis.

**Keywords:** Natural selection, survival of the fittest, Standard Evolutionary Theory, Extended Evolutionary Synthesis, Inevitable Evolution Theory, communicative interactions, altruism, kin selection, inclusive fitness, cooperation, reciprocity, greenbeards, quorum sensing, electrical communication, density-dependent variation, plasticity, phenotypic plasticity, epigenetics, behavioural patterns, aposematism

This (micro-) function is concerned with communicative interactions within the context of evolution. The communicative interaction comes into action in terms of relationships and interactions among a set of agents/components. Thus, this function is concerned with the (evolutionary optimisation of) relationships between individual agents.

**Principle**: Survival of the fittest individual
**Mechanism**: Communicative interactions guiding natural selection[7] among kin
**Emergent pro**perty: Selected phenotypes

"Selection is the differential success within a period, such as a behavioural episode or a generation. Transmission is the fidelity by which selected traits are transmitted to the future, the heritability. ... ... and heritability determines the fraction of selective change that is transmitted to the future." says Frank ([60] *page 1169*).

The role of selection is to promote certain advantageous parts of the gene pool, be it alleles, phenotype traits or a full phenotype, based on an individual's ability to survive and generate surviving progeny that then carries the advantageous genes.

'Natural selection' is an iterative and incremental elimination process by which heritable phenotypic characteristics gradually become more or less frequent in a population as a consequence of individuals interacting with other individuals within their environment resulting in an individual's reproductive success.

Different variants of selection follow the same basic selection mechanisms, but are slightly different, characterised by their effect. Characteristic is that it tends to be local conditions that promote the one or the other of these types; for instance, boldness may

[7] Note the definition of 'natural selection' in Section 1.2.

be an advantage in an environment with scarce food resources, but a disadvantage in an environment with many predators.

'Directional selection' is a kind of natural selection in which a given phenotype is favoured at the expense of other phenotypes, which are therefore de-selected as less fit for the purpose. This means that the genotype/epigenotype behind such advantageous traits is favoured in the population and hence will increase in frequency. There may be many reasons for a lesser fitness: for instance, temporary illness makes an individual an easy target for predators, environmental changes like famine favours the robust phenotype or the phenotype best capable of adapting and finding other solutions to a resource need, or simply that one phenotype is stronger or more robust than another for the conditions at stake; and so on. Directional selection is in particular relevant at environmental changes or when an individual migrates from one place to another.

| **Inevitable Evolution Theory (IET)** | **Standard Evolutionary Theory (SET)** | **Extended Evolutionary Synthesis (EES)** |
|---|---|---|
| Inter-individual interactive behaviour/dynamics, regulated through density-dependent (competitive) interaction:<br>• Quorum sensing<br>• Electrical signalling<br>• Frequency-dependent competitive interaction | Inter-individual interactive behaviour/dynamics, regulated through individual interactive behaviours, kin selection, including:<br>• Altruism<br>• Greenbeards<br>• Cooperation & Reciprocity<br>• Aposematism<br>• Prokaryotic adaptive immunity system | Inter-individual interactive behaviour/dynamics, - adaptive processes regulated through mutual interactions and communicative interaction with the environment:<br>• Plasticity<br>○ Epigenetic foundation for behavioural patterns<br>• Prokaryotic adaptive immunity system |

## 5.1 *Inter-Individual Dynamics, 'Inevitable Evolution Theory'*

The basic mechanism behind evolution in this perspective and micro-function is a) actual exchange of information between individuals in terms of quorum sensing, but also b) frequency-dependent 'competitive interaction', where the impact on evolution comes from physical interactions.

### 5.1.1 *Communication*

#### 5.1.1.1 *Quorum Sensing, Chemical Signalling*

'Quorum Sensing' is a communication mechanism between kin by which exchange of information is achieved through chemical molecular traces or for instance visual signs. Such chemical signalling is constitutive and is received through specific receptors that when the signalling molecule is present (or present above a certain threshold) initiate a chain of events in the recipient, such as transcription of specific genes. Where there are few individuals the secreted signal evades by diffusion /dilution but as the signals increase – for instance by an increased number of individuals in the vicinity or repeated actions leaving traces – the amount of signalling may induce a positive feedback loop and at some point the stimuli increases above a threshold and thereby trigger a response action.

There are many meanings of the concept of 'quorum sensing' / 'quorum response', as summarised in [61], ranging from the one used for human decision making "... a quorum is the minimum number of participants required at a meeting before any officially binding collective decisions can be made ...", over one for animals where quorum is "used ... for the minimum number of committed individuals (i.e. "votes") for a given option that will swiftly trigger concordant behaviour in the rest of the group (i.e. a "quorum response"); (*refs*).", to flocks of birds "the minimum number of closest neighbors a focal individual must monitor in order to be able to satisfactorily match its own movements to that of its group (*ref.*)", and to microbes where "quorum ... translates ... to a threshold level of stimulus beyond which a standard response is effected." ([61] *pages 5-6*).

Dandekar and co-workers in [62] report from their studies on bacterial quorum sensing and the mechanism behind metabolic incentives to cooperate: the chemical signalling in *Pseudomonas aeruginosa* is complex and involves multiple signals and receptors. This communicative interaction controls the production of public goods (secreted extracellular factors like proteases). Not all individuals respond to the quorum signal to produce public goods – that is, a kind of 'cheaters'. Some intracellular enzymes are also controlled by quorum sensing (private goods), and the authors show that such private goods can put a metabolic constraint on social cheating (suppress individuals with the cheating mutant) and thereby a group of cooperators can prevent a 'tragedy of the common' (collapse of the entire colony).

Quorum sensing and response is in particular used for communication to achieve coordination in connection with collective decision making; therefore, see also details in Section 7.1.

#### *5.1.1.2 Electrical Communication*

Communication based on electrical signals is well known from studies of the brain. But now it has been demonstrated that *Bacillus subtilis* in bio-films use the ion channels also for long-range communication; see [63]. Specifically, this bacteria use the potassium ion channel, called YugO, for spatial propagation within the community of bacteria in a biofilm of signals on its metabolic state. Potassium flux, which keeps the internal cellular concentration 40 times higher than in the surrounding environment, is known to be regulated in an oscillating way as a function of the metabolic state of the cell. Their study points to a mechanisms "where metabolically stressed cells release intracellular potassium, and the resulting elevated extracellular potassium imposes further metabolic stress onto neighbouring cells" ([63] *page 61*). The potassium-mediated depolarisation in the neighbouring cells transiently disturbs the forces that maintain the potential across their cellular membrane, and thereby – in a chain of metabolic responses that the authors nicely demonstrate – may give account of the link between the potassium-mediated electrical signalling and metabolic stress and how it propagates in an active fashion. The net result is that the metabolism is coordinated throughout the biofilm.

### *5.1.2 Frequency-Dependent Competitive Interaction*

"In the natural world, chance events (amount of food, weather, etc.) interact with more deterministic biological rules to generate the emergent behaviour of population dynamics." ([64] *page 1176*).

Kapeller and co-workers used Witting's simulation model from [22] at empirical application of population dynamics; see [65]. The authors used it for modelling the selection pressure of density-dependent competitive interactions in a discrete spatio-temporal model. They conclude that Witting's theory is recommendable for spatio-temporal population dispersal models, provided the assumption of the two following factors can plausibly be viewed as the major determinants of population dynamics: a) intraspecific competition and b) density-dependent variation in population growth rate. They summarise the most important aspects of Witting's model this way in [65] (*page 1294*): a) there is no constant reproduction rate; b) the modelling takes into account also data from the previous generation; moreover, c) high population density favours competitive traits and a low population growth rate, whereas a low population density causes a high growth rate, due to a higher environmental capacity and a shift to increased reproductive output.

### *5.2 Kin Selection, 'Standard Evolutionary Theory'*

The basic mechanism behind evolution in this perspective is 'kin selection', based on altruism and similar collaborative/cooperative bilateral interactions.

Webster and Ward in their review conclude "...Personality can affect interactions among individuals within groups, determining social network structure, as well as influencing individual propensity to lead or follow, and produce or scrounge." ([66] *page 771*). The authors reference a series of studies on actual behaviours influenced by personality, ranging from reproductive success, dispersal, response to environmental perturbations, interspecific interactions and competition, and divergence in habitat usage and resource polymorphism. All of this is part of the communicative interaction between kin, and which contribute to determining the actual outcome of the interaction.

#### *5.2.1 Altruism and Inclusive Fitness*

The defining feature of the kin selection theory is the concept of 'inclusive fitness'.

'Altruism' in the context of evolutionary theories refers to a behavioural pattern that increases the fitness of a recipient individual (the beneficiary) at the expense of the fitness of the benefactor (by lowering its relative fitness), whether or not this act is performed with conscious intention of helping another. 'Fitness' denotes an individual's ability to both survive and reproduce, and hence to contribute to the gene pool of the next generation of individuals. The altruistic mechanism operates by helping a kin to gain a higher reproductive success and/or survival, and works only if the beneficiary and benefactor are genetically related. Such relatedness is defined as the probability of sharing a gene ([67]).

Hamilton suggested that the 'Inclusive Fitness' offers an explanation of the evolution of altruism. The underlying assumption is that certain behavioural patterns are affected by genetic mechanisms and hence can evolve by natural selection through favouring those individuals that behave in ways promoting their own fitness or that of related individuals. 'Kin selection' is the mechanism favouring the reproductive success of a being's relatives at the expense of his/her/its own survival and reproduction through an altruistic behaviour.

Hamilton discussed how inclusive fitness should be considered as the fundamental process that embraces kin selection, group selection, and other approaches to social interaction between genetically similar individuals ([60]). Thereby, the Inclusive

Fitness Theory would account for fitness effects to express the evolutionary mechanisms in socially-driven evolution: "Thus, the inclusive fitness of a particular behavioural act is the indirect reproductive gain through the recipient, *B*, multiplied by the relatedness, *r*, minus the loss in direct reproduction, *C*. The relatedness, *r*, measures the genetic discount of substituting the reproduction of the recipient in place of the actor." ([60] *page 1153*). Following Hamilton's suggestion in 1970 and his own use of this concept to explore evolutionary causal relations, his theory has been studied quantitatively and has shown its value/validity in many respects and in numerous studies. It has, however, also been the object of extensive and quite heated disputes (see for instance [68]) – not because of his theory *per se*, but of some derived hypotheses, applications and suggestions.

Frank's review, [60], includes studies addressing whether altruistic behaviour by one species toward a second species (called interspecies altruism) can increase by selection, and concludes that inclusive fitness in the traditional interpretation has no meaning in relation to altruism between species, while Hamilton's rule may be applied when the concepts are interpreted in a broader manner – with problems, for instance, on how to interpret the concept of genetic relatedness. Wyatt and co-workers in [69] analyse the issue of whether interspecies altruism can evolve by natural selection, and conclude with the statement "… our analysis supports Darwin's suggestion that natural selection does not favour traits that provide benefits exclusively to individuals of other species." ([69] *page 1854*).

For the purpose of the present modelling, a slightly different perspective is needed than that of Frank in [60], and others, on the evolutionary theories and models, namely looking for the mechanisms rather than the emphasis on an equation's ability to explore causal decompositions and relations. Nevertheless, changes in phenotypes cause changes in fitness, and with his definition one sees that the behavioural altruistic act induces a disadvantage to the benefactor but gives a reproductive gain for the beneficiary, and the balance in magnitude of these two effects determine the overall reproductive success of that gene pool. "The basic principles of kin selection theory and its descendant ideas always hold. Those principles are: cost and benefits of phenotypes matter; statistical associations between actors and recipients of behaviours matter; and heritability traced from the expression of phenotypes to representation among descendants matters." ([60] *page 1160*), in a context-dependent manner. However, one problem with these simulation studies is that to achieve the simplicity necessary for simulation studies, constrictive assumptions are made, dynamics is ignored as well as certain details of genetics and the developmental complexities that lead the development from genotype to phenotype. For instance, Benton and co-workers discuss the effects of plasticity, trade-offs and inter-generational effects (for plasticity effects, see Section 5.3.2; for timing effect on such dynamics, see Section 8.3.3.2); for instance:

- Changing resource availability (e.g. abiotic conditions caused by the weather, or biotic interactions resulting from density-dependent competition) may trade-off the different life-cycle elements under different conditions: "Under low food conditions, hatchlings from large eggs 'defend' survival at the expense of growth, under high food conditions they defend fecundity, presumably by investing in reserves, at the expense of growth, and in medium food conditions they grow fast." ([64] *page 1174*).
- "One particularly important context-dependent trade-off is the parental provisioning of offspring…" ([64] *page 1174*), as this constitutes the link between

generational conditions. Moreover, the number of off-spring may change with conditions, as may the investment in individual offspring.

Consequently, as different traits will be affected by competition at different times and different densities (both within and between age classes) and under different conditions population dynamics becomes indeed very complex – thus, details for the individual matters.

Altruism affects evolution within a population of individuals of a given species, as shown by a large number of research groups. Altruism is expressed with a vast number of diverse and specific strategies for helping others ([70]). Based on simulation studies with a handful of assumptions (that the authors point out needs to be explored), van Dyken and Wade identify at least four distinct types of altruism differing in the parameters that they modify and which control fitness, and hence these types also differ in their consequential selection pressures ([70]):

- " "Survival altruism" includes common altruistic traits such as defence against predators or parasites, alarm calling to warn of danger …, nest climate control …, and collective thermoregulation" and "… local competition (high crowding, scarce resources) impedes survival altruism." ([70] *page 2489*);
- " "Fecundity altruism" occurs when help donated by an actor causes recipients to increase their investment in reproductive effort, …" ([70] *page 2489*) and "…fecundity altruism is favoured in elastic environments (low crowding and/or abundant resources)…" ([70] *page 2490*), where fecundity is the actual reproductive rate (fertility);
- " "Resource-enhancement" helping occurs when altruists act to increase local resource concentration." ([70] *page 2490*), such as agriculture and rearing of livestock, the latter exemplified by ants milking aphids for their honeydew. This type of altruism is referred to the section on 'Integration' (Section 10.1), since it is realized by individuals increasing fitness through intentional and explicit modification of their environment;
- " "Resource-efficiency" helping enhances the efficiency with which social partners convert resources into fitness. Communication of the location and quality of food source to colony members …" ([70] *page 2491*), for instance through pheromone trails (ants) or dancing (honeybees), or pack hunting (e.g. lions).

Moreover, van Dyken and co-workers find that the first two types of altruism increase the growth rate of social groups and are counter selected by intense local resource competition, whereas the last two increase the growth yield of social groups and are favoured when local resource competition is most intense ([70], [71]).

Also Lehmann & Keller provide a classification scheme for altruistic models, dividing them into kin selection models and greenbeards (see this concept later); however, not as convincing a classification as the above ([72]).

Lehmann and Keller outline four general scenarios that they have identified and where helping (altruism) is favoured and hence will evolve ([72]): 1) the helping provides direct benefits to the benefactor that outweighs the cost of helping; 2) the benefactor is able to alter the behavioural response of its beneficiary by helping and thereby in return receiving benefits that outweigh the cost of helping (i.e. reciprocity); 3) the benefactor interacts and provides help to genetically related individuals (kin selection); and 4) (a special case of scenario 3) related to the greenbeard effect (see Section 5.2.3).

Where there are altruistic individuals, there may also be cheaters benefitting at the expense of others. According to [70], there may – similar to altruism – be diversity in the form of selfish cheating strategies. When kin selection is weak or resources abundant, cheating can prevail, because the selection favouring resource-based altruism is weak. Cheating is dealt with in more detail in Section 6.2.1.1.

### *5.2.2 Cooperation*

'Cooperation' in the context of evolution refers to the act of several individuals taking actions together for mutual benefit or for the common benefit. Thus, an individual engaging in cooperation helps transmitting his/her/its genes to future offspring either indirectly through a relative belonging to the same gene pool, or directly through his/her own increased progeny. Nowak in [67] uses a broader interpretation of the concept of 'cooperation' in his paper, in that it covers the range from kin selection, over three types of reciprocity to group selection. The present author sees 'kin selection' as a unilateral communicative interaction and reserve 'cooperation' to multilateral interactions between individuals, including bilateral interactions. Similarly, group selection is seen as a distinct concept different from cooperation, as a group may consist only of altruists or defectors without cooperation.

Parts of behaviour (heritable cooperative tendencies, which means that individuals interact non-randomly) is correlated with the genotype, and that behaviour is a determinant factor at altruism and cooperation. The non-random element in cooperation can be facilitated by a series of mechanisms "including kin recognition, cognitive bookkeeping (*ref*), spatial assortment with limited dispersal (*refs*), or goal-directed movement away from free riders (*refs*). ..., once established, cooperation can be enforced by social institutions such as direct reciprocity (ref), indirect reciprocity *(ref)*, reputational exclusion (*ref*), and punishment (*refs*)." ([73] *page 247*).

Inclusive fitness theory predicts that individuals will only invest in helping others when they either receive indirect benefits (by helping relatives) or direct benefits from the beneficiary. Even so, "...consistent individual variation in cooperative behaviour is apparently widespread in nature." ([74] *page 2752*), where the authors use the term 'cooperation' in a broad sense. Bergmüller and co-workers in [74] reviews individual differences in animal behaviour for a very large number of species, from invertebrates (such as microbes, Cnidaria and insects) to vertebrates (like fish, birds and mammals), demonstrating a large variety of cooperative phenomena. They say that "... assuming the existence of an optimal behavioural phenotype, natural selection should reduce genotypic variation over time (*ref*). However, behavioural phenotypes typically show heritable variation, which appears not to be eroded by selection ..." ([74] *page 2751*). Further, they say that "Animal personality has been found to be heritable ... (*refs*) and to affect fitness ... (*refs*) showing that it is subject to evolutionary change." ([74] *page 2751*). Among others, from their paper one identifies the following mechanisms for establishing and maintaining cooperation: a) consistency in the level or type of cooperative behaviour provide a fitness gain; b) group level benefits from cooperative task sharing seem to increase reproductive output for cooperative breeding; c) commitment as a plausible explanation for consistency in cooperative behaviour;

consistency in behaviour may be a means to reduce conflict; and d) stabilising cooperation through punishment mechanisms.

Lehmann and Keller say that there are two general scenarios where helping can evolve and the act is cooperative ([72]): a) when there are direct benefits for both, and the evolutionary pressure on helping is expected to be high when the fitness of an individual critically depends on its investment in cooperation; and b) at repeated interaction, provided that the interacting individuals have an initial tendency to be cooperative.

Lehmann and Keller outline different mechanisms involved in facilitating helping in general, therein cooperation, such as ([72]): 1) evaluation of the cooperative tendency of the counterpart; 2) use of spatial cues with individuals expressing conditional altruism in the natal nest or colony (common in social insects); 3) comparison of phenotypic characteristics with those of other individuals; and 4) alteration of the interaction with coercion, punishment, and policing of defectors as a mechanism for suppressing selfish behaviour. They further suggest four types of models as particularly interesting principles for the evolution of cooperation and altruism ([72] *pages 1372-3*): i) 'spatial structuring models', where the actual selective force is kin selection; ii) 'reproductive skew models', defining the conditions under which the best strategy is to cooperate and sacrifice part or all of its direct offspring production; iii) 'tag recognition models', which are in essence greenbeard models (see next section); and iv) 'group selection models' as a multi-level selection approach to partition selection into components within and between groups. See more on type (iv) under the section 'Group Selection (Multi-Level Selection)'.

### *5.2.3 Greenbeards*

"A greenbeard gene is defined as a gene that causes a phenotypic effect … that allows the bearer of this feature to recognize it in other individuals, and results in the bearer to behave differently toward other individuals depending on whether or not they possess the feature." ([72] *page 1370*), whether this is intra- or interspecies.

Required is either a single gene or tightly linked genes encoding both the cooperative behaviour and causing cooperators to associate. "Greenbeards are one of the two ways in which natural selection can favour altruistic behaviour, with the other being interactions with genealogical kin …" ([75] *page 1344*), however, greenbeards are only favoured in the case that its frequency exceeds a certain value, so it is obvious that it needs cooperative mechanisms in order to gain importance in an evolutionary context.

### *5.2.4 Reciprocity*

'Reciprocity' refers to repeated encounters between individuals, which may comprise different species ([67]). Like 'cooperation', 'reciprocity' constitutes bi- or multilateral communicative interactions.

According to Nowak, 'direct reciprocity' requires repeated encounters between the same two individuals, and both shall provide help to the other, which is less costly for the benefactor than it is beneficial for the beneficiary; the mechanism with which it affects evolution is kin selection. 'Indirect reciprocity' comparably relies on asymmetrical interaction between more than two individuals, relying on accumulated reputation based on previous deeds in similar situations ([67]). Natural selection favours strategies that base the decision to help on the reputation of the recipient, and both theoretical and empirical studies of indirect reciprocity show that helpful people are more likely to receive help themselves ([67]). A third type of reciprocity is 'network reciprocity', which is a type of indirect reciprocity yet multilateral in its nature, as everybody in principle interacts equally likely with everybody else, and hence beneficial help to someone will sooner or later return to the benefactor.

The cognitive element in the indirect and network reciprocity may explain why it appears that only humans seem to engage in the full complexity of indirect reciprocity; if so, natural selection will favour further evolution of the cognitive aspect.

In a recent study by Hein and co-workers ([76]), it is nicely demonstrated by means of functional magnetic resonance that altruism and reciprocal altruism are indeed distinct behavioural actions with each their behind motivation and mechanisms: "Empathy-based altruism is primarily characterised by a positive connectivity from the anterior cingulate cortex (ACC) to the anterior insula (AI), whereas reciprocity-based altruism additionally invokes strong positive connectivity from the AI to the ventral striatum." ([76] *page 1074*). Further, they demonstrate that "predominantly selfish individuals show distinct functional architectures compared to altruists, and they only increase altruistic behaviour in response to empathy inductions, but not reciprocity inductions." ([76] *page 1074*).

### *5.2.5 Aposematism*

Aposematism is a signalling mechanism for avoiding becoming a prey. It is defined as the "... combination of unprofitability (e.g. physical protection, chemical toxicity, or difficulty of capture) with one or more signals (such as warning or conspicuous coloration) warning of that unprofitability to potential predators." ([77] *page 933*). This is a kind of cheating or deceit, as one 'pretends' to be something / someone else than one is, while it is simply exploitation through mimicking of the signalling mechanisms, such as warning colours or patterns that signals danger, that have been established and hence widely incorporated in interaction between predators and potential prey.

In the case of signalling by means of warning colouring, it was speculated for a long time how this would be able to evolve in nature, simply because being brightly coloured means that one is highly visible to predators, which is usually a selective disadvantage for prey. However, it turns out that neophobia and dietary conservatism are strong preferences in the foraging of many predators ([77]), avian as well as others, and thereby often paradoxically serve as a selective advantage for the prey.

According to Marples and co-workers ([77]), "The repeated evolution of aposematism across and within a range of taxonomic groups constitutes strong evidence that aposematism can readily evolve in a wide range of ecological contexts and predator-prey systems." ([77] *page 937*). The authors list conditions identified to avoid immediate demise for a fully palatable conspicuous new prey morph: 1) both the old and new morphs are fully palatable; 2) the new morph is much more conspicuous than the old; 3) the new colour is associated with existing aposematic signals; 4) irrespective of whether one or several predators have access; 5) the novel prey exist over several generations and in increasing numbers; and 6) the novel prey change from being a minority to being the majority, and hence become fixated.

### *5.2.6 Prokaryotic Adaptive Immunity System*

The CRISPR-Cas adaptive immune system – also described in Sections 5.3.3 and 9.3.3 – seems to exist in several versions that range from Neo-Darwinian at the one extreme to Neo-Lamarckian at the other extreme ([78]); or at least the various versions each have characteristics that better fit the one or the other of the evolutionary perspectives, and seemingly in a continuum between the two extremes. It is the Neo-Darwinian version that is described here.

The immunity system is applied by prokaryotic organisms infected with bacterial and archaeal organisms. The CRISPR-Cas immune response as a first stage (adaptation stage) involves the insertion of pieces of foreign DNA, such as a viral or plasmid genome, specifically into the CRISPR array; these inserts are denoted spacers as opposed to the sequences in the foreign DNA that give rise to spacers, called protospacers. Next stage is the utilization of processed CRISPR transcript (crRNA) as guides for inactivation of the cognate target. "The net result is the acquired, heritable, highly specific and efficient protection against the cognate (parasitic) element." ([78] *page 2*).

The critical point is the self-nonself discrimination, and such discrimination is the feature for distinguishing whether a CRISPR-Cas mechanism belongs here or in the parallel section for the Extended Evolutionary Synthesis. Inability of self-nonself discrimination leads to an autoimmune reactions, resulting in cell death from suicide. It has turned out that the type II-A CRISPR-Cas system (from *Streptococcus thermophiles*) inserts an apparent random spacer, resulting in cellular suicide, except that "the few that incorporate spacers homologous to the invader genome could survive" ([78] *page 3*), and therefore, this type is extremely wasteful but will nevertheless continue to exist.

### *5.3 Regulatory Adaptation, 'Extended Evolutionary Synthesis'*

The basic mechanism behind evolution in this perspective is 'adaptation', based on adaptive interactions with the environment, enabled through plasticity and modularity.

Laland and co-workers in [3] emphasise that there are multiple mechanisms contributing to the development of an organisation rather than only through the transmission of DNA. Throughout the development, there are multiple kinds of interactions that are now known to contribute to the shaping of the phenotype, from (bio-)chemical interactions (nutrients, hormones, polluents), over behavioural interactions (maternal / parental care and teaching or imitation; see later under 'Life-history Theory'), to inheritance of symbionts. This is a broader notion of inheritance, [3], and "... can bias the expression and retention of environmentally induced phenotypes, thereby influencing the rate and direction of evolution [*ref*]." ([3] *page 4*).

#### *5.3.1 The Genetic/Epigenetic Foundation of Behavioural Patterns*

Wolf & Weissing say that "In many animal species, individuals of the same sex, age and size differ consistently in whole suites of correlated behavioural tendencies, comparable to human personalities (*refs*)." ([79] *page 3959*). So, what is it that determines behavioural patterns?

One of the latest news is that even primitive life forms like sea anemones – and others at the same primitive developmental stage with a simple nervous system – have 'personality', where "'Animal behaviour' means that individuals differ from one another in either single behaviours or suites of related behaviours in a way that is consistent over time." ([80] *page 1*), and that such behaviour is correlated with relatedness ([81]). Wolf and Weissing emphasize that "the concept of personalities does not require that individuals are completely consistent in their behaviour but rather that individual differences are consistently maintained over time and across contexts (*ref*)." ([79] *page 3959*).

Observations similar to those of the sea anemones are found also for the cockroach *Periplaneta americana* ([82]). This shows that also primitive organisms have a consistent behaviour. Such consistency in behaviours is the background for the effect leading to altruism and cooperation in more advanced organisms than those mentioned, and thereby behavioural patterns may become a fitness variable in the selection process.

Behavioural types "often exhibit (i) time consistency of behaviour (...), and (ii) suites of correlated behavioural traits" ([79] *page 3960*). Further, the authors say that behavioural correlations can often be understood in terms of the genetic, physiological, neurobiological and cognitive systems underlying behaviour, "examples include pleiotropic genes (*ref*), hormones (*ref*), neurotransmitters (*ref*) and emotions (*ref*) affecting multiple traits at the same time." ([79] *page 3960*). The authors express that behavioural correlations caused by inherently stable states[8] may reflect either a genetic polymorphism or phenotypic plasticity. For labile states (which "... include gene expression patterns, levels and compositions of hormones and neurochemicals, receptor sensitivity and density, blood pressure, energy reserves, ..." ([79] *page 3961*)) in some situations, the state and behaviour of individuals are coupled with a positive feedback mechanism, which can lead to consistency in labile states and associated state-dependent behaviour and hence stabilisation of the behaviour. For instance, "Animals often learn how to recognize predators (*ref*), which in turn reduces the cost of exploring and foraging in a risky habitat." ([79] *page 3961*), an adaptive pattern that increases fitness within the current context. The authors conclude in this respect that initial variation in states or behaviour combined with positive feedback mechanisms can explain adaptive behavioural consistency and "can explain that seemingly minor and labile differences in state are enhanced into major and stable differences (*refs*)." ([79] *page 3966*).

In section 5.2.4, it was shown that behavioural patterns are imprinted in the structural architecture of the brain; at least this was shown for humans.

The human psychology, normal and pathological, is pretty complex. There are behavioural patterns among the 'normals' (humans); for instance, a psychological tool like the enneagram include 9 different personality traits that sometimes exist in combinations. Other tools show different sets of behavioural categories. Orthogonal to this come highly sensitive traits (see e.g. the review [84], which by the way conclude that such variance in sensitivity is also found in animals), and so on, and so on. It is anticipated that all of the behavioural patterns may influence for instance the kin selection mechanism and the mechanisms in all of the functions following the present function (i.e. Chapters 6 to10). The behavioural patterns may determine the individual's influence on group fitness or fitness in a cultural context – see these concepts later. Even if there are many more factors affecting evolution for the individual, for the present purpose it suffice to outline only the mechanisms behind the effects of altruism and cooperation on an individual's fitness and hence their ability to promote the dispersion of their gene pool into the future either directly by themselves or indirectly through genetically related kin individuals.

---

[8] The authors define a state of an animal as: "all those features that are strategically relevant, i.e. features that should be taken into consideration in the behavioural decision in order to increase fitness (*refs*)." ([79] *page 3960*), or in a later reference of the same authors: "those features of an animal (e.g. morphological, physiological, neurobiological or environmental) that affect the cost and benefits of its behavioural actions and thus its optimal behaviour [*refs*]." ([83] *page 440*).

Wolf and Weissing in [79] further discuss how different behavioural types can adaptively coexist within a single population in a set of scenarios, as follows:

1) Externally induced differences in states can coexist even without achieving the same fitness: "Different behavioural types can adaptively coexist whenever individuals differ in state and behavioural variation among types reflects a state-dependent response of individuals (*refs*)." ([79] *page 3963*); here, consider the life-history theory discussed in Section 8.3.3.1. The behavioural variation in this context does not reflect a genetic polymorphism, but may be explained by phenotypic plasticity ([79]);
2) Frequency-dependent selection; negative frequency-dependent selection can in a producer scrounger scenario lead to fitness equality through selection, because the rarest behavioural types has an advantage until there is a balance in fitness. The authors list three general mechanisms that at negative frequency-dependent selection can give rare behavioural types an advantage over more common types: i) competition avoidance; ii) enemy avoidance, and iii) complementation. The authors emphasise that variation in behaviour caused by frequency-dependent selection may be cause by either phenotypic plasticity or genotypic polymorphism;
3) Spatial variation in the environment, where they argue that context-dependent fitness is the rule rather than the exception. If individuals are constrained in their ability to match the environmental conditions the variation will be maintained for the population; this variation can in principle be realized by behavioural plasticity or a genetic polymorphism; see [79]. According to the authors, the result will arise as a consequence of phenotypic plasticity, and there will not be fitness equality;
4) Non-equilibrium dynamics – despite sustained variation in fitness. "In several examples, it has been demonstrated that non-equilibrium conditions have a high potential for maintaining variation even in cases where equilibrium theory would predict the dominance of a single behavioural type (*refs*)." ([79] *page 3965*). The authors state that phenotypic variation in this respect may or may not be associated with genetic variation.

Further, Wolf and co-workers – in a subsequent simulation study ([83]) – conclude "Whenever sufficient variation among individuals is present, however, a coevolutionary process between responsiveness and consistency is triggered which, in turn, gives rise to populations in which responsive individuals coexist with unresponsive individuals who show high levels of adaptive consistency in their behaviour." ([83] *page 447*). Further, in the absence of sufficient initial variation, "… individuals evolve a mixed strategy which gives rise to inconsistent behaviour in repeated social interactions." ([83] *page 447*).

### *5.3.2 Plasticity*

See also the discussion on 'plasticity' in Section 4.3.2. Here, the aspects of plasticity related to its role as the mechanism behind adaptability are addressed – that is, as a response to the communicative interaction with the environment. For instance, Beldade and co-workers in their invited review sum up that "… DNA methylation plays a key role in mediating many cases of environmentally induced phenotypic variation (*ref*) including caste determination in honeybees (*ref*). … leading to the suggestion of an association between methylation and morphological specialization (*ref*).” ([33] *page 1357*). Benton and co-workers add to this that plasticity is likely ubiquitous, but that the traits affected will vary between organisms (see [64]), for instance, some organisms have fixed size or age at maturity, but that the plasticity instead may show up in other traits like reproduction or senescence; and further, they say that traits are unlikely to be independent of each other, but may co-vary in a positive or negative fashion.

Environmental cues can have systemic effects but also localised effects in developing organisms. In this process, phenotypic plasticity – also called developmental plasticity – increases the phenotype’s ability to adjust to environmental conditions, usually as a result of induced changes in gene expression. Such adaptation may be implemented not only through DNA methylation (which is the one typically mentioned), but also by means of regulatory microRNAs and post-translational modification of regulatory proteins, as well as mechanisms of signal reception and signal transduction ([33]). Environmental cues can invoke direct biochemical effects and/or be mediated by the neuroendocrine system through its natural role of transducing a trigger signal into a physiological response – that is, a developmental trajectory. “a hormonal regulation has been characterised for most, if not all, well-described examples of developmental plasticity (*refs*). … … often associated with different sensitivity thresholds (*ref*) and/or different sensitivity periods (*refs*).” ([33] *page 1356*). Such trigger may result in either a change in hormonal titre and/or even at the level of dynamics of the hormone production, since some hormones regulate the production or secretion of other hormones.

The environmental prompts are transduced into cellular ones and propagated further by means of hormones, metabolites, receptor molecules, nerve signals, osmotic changes or physical interactions among cells ([55] *page 2708*).

Further, Moczek and co-workers in [55] (*Figure 1*) convey the following examples: 1) “When a bluehead wrasse … male … is removed from his harem, a female … will change phenotype completely and become a male.”; and 2) the Arctic fox’s seasonal shift in coat colour. These examples show that plasticity is not only taking place in the developing organism, but also in adult individuals.

Schlichting and Wund reference a couple of studies explaining the mechanism(s) behind plasticity’s facilitation of evolutionary responses by means of modelling investigations: 1) “through initial evolution of adaptive plasticity followed by fixation of a new phenotypic optimum.”; 2) “by blunting the effective size of the environmental change and subsequently increasing the rate of adaptation.”; 3) by “ameliorate the effects of the flow of maladaptive alleles from central populations by raising fitness and, concomitantly, overall population size…” ([48] *page 658*). What happens is that the ability to respond to environmental changes “…enables the generation of functional, integrated phenotypes, despite development occurring in previously unencountered, or greatly altered, conditions. … Developmental plasticity thus has the potential to determine which phenotypic and genetic variants become visible to selection in a novel environment, thus delineating the nature and magnitude of possible evolutionary responses.” ([39] *page 2*).

Snell-Rood and co-workers in [56] introduce the concept of ‘relaxed selection’, referring to the weakening of the selection resulting from the ability to adapt to changing conditions. The weakening of selection due to environment-specific gene expression (the nature of plasticity) enables mutations (in principle deleterious as well as beneficial ones) to accumulate in the population, and this to a larger extent than for constitutively expressed genes. Thereby, the plasticity as an evolutionary factor enables a larger variation within the population, and at the same time affects the purification and the fixing of mutations, because of the delayed selection pressure on them. By this mechanism, it has been shown that sporulation, a complex response to stress in bacteria, is predominantly lost through neutral processes of mutation accumulation instead of selection, when the populations are not induced to sporulate for 6000 generations ([56]). Nevertheless, when we think of human evolution, 6000 generations comprise quite a lot of calendar time where one cannot expect constant environmental conditions; and still it is a lot less than the span of existence of humans.

As said previously, also Sharov in [49] refers the mechanisms behind plasticity to the domain of epigenetics. He divides the overall mechanisms of plasticity into four groups, of which two are dealt with in Section 4.3.2; the other two are:

- Connection is the ability to coordinate changes in previously independent components through communicative interactions, depending on the establishment of an agent that is able to perform the link. An example is connection between regulatory networks resulting in coordinated variation of cell organelles or organism organs
- Interaction is the ability of organisms to communicate and coordinate their activities. An example is organisms living in colonies (polyps, ants rodents), where functions of individuals are regulated by contact or quorum signalling

In summary, environmental cues can have local or systemic effects in developing organisms and/or in adult organisms, modifying the fitness and hence, the selective pressure at evolution.

### *5.3.3 Prokaryotic Adaptive Immunity System*

"The CRISPR-Cas system of prokaryotic adaptive immunity displays features of a mechanism for directional, Lamarckian evolution. Indeed, this system modifies a specific locus in a bacterial or archaeal genome by inserting a piece of foreign DNA into a CRISPR array which results in acquired, heritable resistance to the cognate selfish element." ([78] *abstract*). The neo-Darwinian version of this system is described in Section 5.2.6, to which is referred with respect to the general mechanism of this immune system.

As said in Section 5.2.6, the distinguishing feature is the self-nonself discrimination, and only discriminatory CRISPR-Cas systems are truly of the Lamarckian type ([78]). For type I-E CRISPR-Cas system, for instance at *E. coli*, the ratio of foreign over host DNA among the inserted spacers is 100-1000 times in favour of the host, and thereby the autoimmunity is avoided ([78]). In particular, the spacers are acquired at stalled replication forks and are produced during the repair of double-stranded breaks associated with stalled replication forks; so this process of generating spacers is not by itself based on an intrinsic difference between the foreign and the host DNA, but rely on a much higher density of replications forks and subsequent also breaks. Koonin and Wolf explains that, after recognizing the cognate protospacer, the Cas machinery efficiently generates new spacers, and apparently without dissociation from the target DNA and without the special mechanisms for recognition of a protospacer adjacent motif (PAM) that other types of CRISPR-Cas systems use in the adaptation phase; see Section 5.2.6. Koonin and Wolf express that "In stark contrast, the type I-E CRISPR-Cas system seems to operate via a bona fide Lamarckian mechanism where the mutational process is dominated by directional, adaptive mutations which is achieved via the coupling of spacer acquisition with replication accompanied by the DSB formation and the priming mechanism." ([78] *page 4*).

Koonin and Wolf further express that "The key feature of the Lamarckian mode is the non-randomness of mutations that is achieved via evolved mechanisms that are highly specific, elaborate and subject to regulation." ([78] *page 5*).

# 6 Stabilisation: Balancing Efficiency and Effectiveness at Evolution

**Abstract.** Function 3, stabilisation (see Table 2 and Table 3 in Chapter 14), is concerned with maximising efficiency ('doing the things right', taking the fullest advantage of evolutionary resources), while founding stability – that is, enhancing the capacity and capability for 'doing the right things right'. Such optimisation may come at a cost somewhere somehow. The principle is 'survival of the fittest group of individual resources', and the mechanism is 'natural multilevel selection', explained for the three perspectives: Standard Evolutionary Theory, Inevitable Evolution Theory and Extended Evolutionary Synthesis.

**Keywords:** Natural selection, survival of the fittest, Standard Evolutionary Theory, Extended Evolutionary Synthesis, Inevitable Evolution Theory, group selection, multi-level selection, group dynamics, meta-population dynamics, flock dynamics, team behaviour, cheating, deception, social plasticity

The definition of 'system' explicitly states that a system has an internal structure. Such internal structure encompasses individual resources and their relationships. Thus, the necessary regulation to achieve optimisation may have the consequence of prioritising that which is beneficial for the system as a whole or for parts of its internal structures at the expense of individual resources, rather than doing what is beneficial for given individual resources. Therefore, here, the concept of 'group' comes in as embracing a set of individuals.

**Principle**: Survival of the fittest group of individual resources
**Mechanism**: Natural multi-level selection[9]
**Emergent Property**: Selected phenotypes

A 'group' comprises interacting individuals, and those individual resources are also individually exposed to the evolutionary selection as described in the previous chapter, so the evolutionary pressure for a group of individuals comes on top and is intertwined with the individual selection – therefore, the term 'multilevel selection'.

Groups in the sense that the literature on evolutionary theories uses this concept are similar to the concept of systems in the sense that groups are self-contained entities with internal structure. Moreover, when thinking of groups in the context of evolution of biological beings, such groups have dynamics, their component parts are inter-relational, acting internally, and also affecting conditions externally. The difference between a system and a group is that a 'group' 1) may dynamically change in various contexts – which a system does not; 2) may have overlapping groups; and 3) there is no requirement regarding discrete functions or roles.

---

[9] Note the definition of 'natural selection' in Section 1.2.

The closest that the author has come to finding a formal definition of 'group' in the literature on evolutionary theories is Luo's "a collection of individuals" ([85] *page 43*) and Pievani's a "group is a context of relationships that make adaptive sense to individual behaviours, in most cases pre-adapted by classical natural selection." ([86] *page 320*). Therefore, a definition of 'group' that serves the present purpose was established, based on the above definitions, the literature's use of the concept, and the definition of 'social group' that was found on Wikipedia (accessed 14th February 2015) together with inspiration from the discussion by Gerkey and Cronk in [87]: A group is a set of individuals comprising a self-contained wholeness (an organisation) with coordinated actions, shared motives and purpose, accepted norms and values with respect to matters relevant to the group. So, the concept of 'group' does not apply for a set of cells organising into a tissue or a multicellular organism.

Gardner points out that ambiguity exist over the precise meaning of group trait and group fitness ([88]). Such ambiguity reveals itself in quantitative modelling. For instance, the change may take place in the frequency of the different types of individuals within the group, or as a change in the frequency of different types of groups. To achieve unambiguity in this respect, he distinguishes between 'unit of selection' (the entity upon which the selection acts), 'arena' (the assemblage within which selection acts), 'character' (the numerical property of the units, whose aggregate change may be driven by selection) and 'target' (the numerical property of the units which provides the measure of success); and then defines 'natural selection' as "a particular kind of selection defined by the conjunction of a particular unit, arena, character and target." ([88] *page 306*) – that is, this definition is different from the definition in Section 1.2; however, by relaxing the constraint ("a particular kind") the difference is minor, and his suggested framework will still work. Further, "That is, the change in average fitness ascribed to the action of natural selection is equal to the (additive) genetic variance in fitness (*ref*)." ([88] *page 307*), implying that change is driven by differential fitness. Also social groups can be considered a viable unit of selection. He defines "... the reproductive success of any unit in terms of its expected long-term genetic contribution to future generations." ([88] *page 310*).

'Group-level traits' are characterised by Smaldino as "The properties that allowed one group to triumph or persist against another in these cases did not belong to each individual group member, but rather emerged from the organized interactions between those individuals." ([73] *page 244*). In this respect, it is also important – as shall be seen in the below – that the individuals and their specific advantageous or detrimental traits are balanced in number and character within a given trait – that is, the intra-species dynamics of a population that may be divided over several patches.

In the below, there are three aspects concerning meta-population dynamics in evolution: i) the dynamics concerning a population of populations (or groups within groups); ii) the concept of 'group' in the most commonly used meaning in the literature on evolution; and iii) patterns of group behaviour. The last (i.e. patterns of group behaviour in general) has a focus on optimisation of the system's behaviour through its members. The meta-population dynamics in general has the purpose of for instance dealing with issues like habitat suitability, competition between kin, inbreeding, and resource competition; finding the right habitat is key to fitness, determined by factors like availability of appropriate food resources, density of inter- and intra-species competitors and/or cooperators, suitability of nesting and breeding options, and/or sheltering as well as density and properties of predators. The three aspects are dealt with in each their perspective, as is seen from the table.

| Inevitable Evolution Theory (IET) | Standard Evolutionary Theory (SET) | Extended Evolutionary Synthesis (EES) |
|---|---|---|
| Meta-population dynamics:<br>• Flock dynamics | Meta-population dynamics:<br>• Group dynamics, multi-level selection<br>• Group behaviours<br>  o Cheating and deception in the context of group dynamics and cooperation | Meta-population dynamics, adaptive processes:<br>• Acquired team behaviours, social plasticity |

## *6.1 Meta-Population Dynamics, 'Inevitable Evolution Theory'*

The basic principle behind evolution for this function in this perspective is related to a population of individuals clustering into groups, with a marked yet steady display of migration among the groups to optimise fitness in a balanced way between the individual and the population as a whole.

With meta-X being 'X-about-X', a 'meta-population' is a population that can be divided into a number of geographically separated subpopulations – that is, meta-population dynamics is concerned with intra-species spatial distribution and the dynamics in this respect. Most populations in nature exhibit some form of spatial structure, for instance in terms of fragmented habitats (like clan/herd territories, cities versus farming areas, mountain areas or deserts) or just simply because dispersal is limited ([89]). When dispersal is limited, individuals aggregate within clusters where related individuals tend to live together, affecting the dynamics of altruism and hence also population dynamics. At the same time there is a fitness cost to the clustering in terms of reduced fecundity or reduced survival because of increased competition ([89]), which may even counteract the beneficial effects of the clustering.

The dynamics include/involve selection at the (sub-)population level. Modelling in this respect has in particular been used to describe meta-population dynamics for insect pests, infections spreading between individual hosts, the ecological stability of species within isolated circumscribed ecology (such as a pond), species that are territorial as well as species at risk of extinction. Therefore, it will also include herd and clan dynamics. A given population is relatively independent of parallel 'sister'-populations, while at the same time migration/dispersal dynamics between groups are relevant for the fitness of individuals within their social context and hence for the survival of both individuals and such subpopulations.

### *6.1.1 Flock Dynamics*

Relevant here is the mechanisms behind characteristics that influence flock dynamics – including behavioural ones, while the influence of decision-making strategies are referred to Chapter 7. In short, it is the stochasticity and specific dynamics that are the factors in flock dynamics and may be stable over ontogeny and/or across situations.

The introductory statement in the review of Krause and co-workers is that "relatively little is known about the size, composition and dynamics of free-ranging fish shoals." ([90] *page 477*), and not much new information has been added since then. A lot of it is about decision-making criteria related to the question of joining a given shoal, where the 'oddity effect' regarding preference toward conspecifics, manoeuvrability, size and colouring are important factors. Beyond the decision-making aspects, the authors reference a couple of factors regarding shoal dynamics: a) "Familiarity among the members of a shoal may reduce the fitness cost of competition by reducing aggression between the contestants." ([90] *page 489*); b) "fathead minnows that originated from the same shoal exhibited more effective antipredatory tactics under predatory threat ..." ([90] *page 489*); c) "that association preferences of Arctic charr are at least partly based on major histocompatibility complex (MHC) genotype, ..." ([90] *page 491*), pointing at a role of kinship; d) the release of 788 marked (free-range) fish from 10 shoals showed that "...distribution of the marked fish between shoals was not different from random suggesting low shoal fidelity which may be a result of breakdown of shoals overnight and/or random reformation in the mornings (*ref*)." ([90] *page 495*); and "In some species, males are territorial and keep harems thus controlling group structure (*ref*)." ([90] *page 496*); "Fish may develop an attachment to familiar sites ... Remaining in the same area will allow an individual to build up an increasing store of information on predator habits and distribution and food locations." ([90] *page 496*).

A recent study by Cavagna and co-workers of the dynamics of natural flocks of birds applies quantum mechanics, using an inertia-spin model according to which the birds communicate their movements to each other in either of two ways ([91]): by spin fluctuations or density fluctuations. The authors demonstrate that the flock has to be either small or large, while medium-sized flocks cannot propagate their information in a linear and underdamped way – that is, "...either under the form of orientational fluctuations or under that of density fluctuations, making it hard for the group to achieve coordination." ([91] *page 1*). The consequence is that information cannot propagate appropriately within medium-size populations, rendering such flocks unviable.

### *6.2 Meta-population Dynamics (Group Selection), 'Standard Evolutionary Theory'*

The basic mechanism behind evolution in this perspective is selection operating on group fitness characteristics, based on behavioural traits of the group and its individuals – that is, multilevel selection.

#### *6.2.1 Multilevel Selection*

The discussion on group selection within the literature is fierce (see for instance [86], [92]), and the fight is still ongoing. The mentioned two references deal with the theoretical and formal premise of the debate, centred on group selection theories. A reason might be that the models discussed are of a descriptive correlational nature rather than a causal nature, even if they attempt to find causal models behind observations made in nature.

At the one extreme (according to [60], [72], [93] and many more), it is wrong to say that there is a dedicated group selection theory, with the argument that the mathematical modelling of groups may be achieved by means of kin selection models that considers individuals – that is, there is no such thing as dedicated group processes (mechanisms) in evolution. Frank concludes that the group structured perspective works like a special case of kin selection within the broader analysis of phenotypes, and that pathways of causation replace the group selection theory.

At the other extreme, several papers successfully develop theories for modelling group selection, for instance: [85], [88], [94], [95]. Simon and co-workers strongly conclude that "…this kind of group selection is not mathematically equivalent to individual-level (kin) selection." ([95] *page 1561*) in a dynamic setting, thereby contrasting the above mentioned reviews of Frank and Lehman and co-workers (and more researchers) by pointing at scenarios that will not be solved by inclusive fitness approaches. van Veelen is more moderate in his statements, saying 1) that different views need not be incompatible; 2) that not all group selection models are the same; and 3) that there are models containing synergies; and 4) "that as soon as a group selection model implies a public goods game that is not linear, inclusive fitness can give the wrong prediction." ([94] *page 594*).

Correspondingly, van Veelen's group selection theory in [94] is based on the principle that selective forces work at the different levels (group level versus individual level) and in opposite directions, with factors like altruism, selfishness, spitefulness and mutualism involved. In a later paper by van Veelen and his co-workers ([92]), the authors take their gloves off in the discussion on group selection versus kin selection and their alleged 'equivalence'. Their discussion boils the problem down to researchers' lack of self-assessment and lack of assessment of assumptions' validity, and/or uncritical adoption of established approaches from the literature. The authors conclude that inclusive fitness gives the correct prediction only for a well-defined strict subset of group selection models with defined characteristics, such as when the dynamics are payoff monotonic and when effects on fitness are additive. They re-analyse a set of cases from the literature and point out what is wrong and how these problems may be remedied. In a recent paper, van Veelen and co-workers in [96] elaborate on the same topic and with the same conclusion. Also Allen and co-workers use surprisingly strong words against inclusive fitness as a universally applicable theory to explain the evolution: "…, it is claimed that inclusive fitness theory *(i)* predicts the direction of allele frequency changes, *(ii)* reveals the reason for these changes, *(iii)* is as general as natural selection, and *(iv)* provides a universal design principle for evolution." ([97] *abstract*). In their paper, the authors evaluate these four claims, and show that all of them are unfounded.

It is not the task of the present study to solve or reconcile their dispute. The papers of van Veelen and others similarly are so convincing from a professional perspective that the discussion will be left here and the attention turned toward the Mereon Matrix for reconciliation to see what it can bring.

The Mereon Matrix's template information model shows that a pluralistic approach likely is the accurate one as said earlier – that is, normally a 'both/and' rather than an 'either/or' is the accurate standpoint. And again, a system has internal structure. So, from the present modelling perspective, based on the understanding synthesized from the literature together with the template model characteristics, the existence of a theory on multilevel selection (group selection) is supported in the present model.

According to Smaldino, "If well-defined groups compete, however, and the variance of a trait tends to be higher between than within groups, then it is theoretically possible for an altruistic, group-beneficial trait to emerge, because groups with many individuals possessing such a trait will have higher mean fitness than groups with fewer altruists." ([73] *page 245*). This characterises the multilevel selection framework.

Simon and co-workers describes 'group selection' as "... is about the effects of group-level events on a two-level evolutionary process." ([95] *page 1562*). The reason that it operates at two levels is that while the mechanism of evolution operates at the group level, it is inescapable that such group-oriented selection mechanism penetrates and thereby interacts with and affects the individual level, directly and/or indirectly.

In nature, many groups have collective behaviours without specialisation. Those with specialisation are discussed in the section on 'patterning' (Chapter 8).

Both Luo and Simon and co-workers present multilevel selection theories that operate on two levels in combination, a group level and an individual level, based on the idea, as described by Simon and co-workers, that "... genes encoding traits that are detrimental to the individual carrying those genes might still thrive on an evolutionary scale if the trait confers an advantage to the group in which the individual is living." ([95] *page 1561*). Thus, group selection is about the effect of group events on individuals in a synergistic two-level evolutionary process: "... the group-level events directly affect group-level population dynamics, and only indirectly affect individual-level population dynamics, and conversely for individual-level events." ([95] *page 1566*). Furthermore, they define that a trait evolves by group selection in a model if it only establishes itself when group-level events are present, and a trait is assisted by group selection in a two-level population dynamics if it only establishes itself more quickly and/or more completely only when group-level events are present – that is, they only talk about group selection when there is a demonstrable effect of such grouping. Luo demonstrates that "selection at the group level is favoured when group-level events occur frequently relative to individual-level events, when there is little or no mutation, and when there are many groups relative to the number of individuals in each group." ([85] *page 41*).

In conclusion, the Mereon Matrix template information model tells us that the system has an internal structure and therefore, selection will act upon all component elements – that is, the group structures as well as the individuals.

##### 6.2.1.1 Cheating and Deception in the Context of Group Dynamics and Cooperation

According to Ghoul and co-workers, cheats are broadly classified on the basis of four distinctions: "(i) whether cooperation is an option; (ii) whether deception is involved; (iii) whether members of the same or different species are cheated; and (iv) whether the cheat is facultative of obligate." ([98] *page 318*).

van Dyken and Wade mention 'cheaters' and 'selfish cheating strategies' and say that cheaters may have as many diverse strategies as altruistic individuals and may create conflict within a group by securing personal gain at the expense of others – the polar opposite of altruism ([70], [71]).

Ostrowski and co-workers studied 'cooperation and conflict' (cheating) in a social amoeba while searching for the corresponding genes ([99]). The genomic signatures were quite complex and consistent with frequency-dependent selection acting to maintain multiple alleles. They suggest that their results indicate stalemate rather than an arms race, with balancing selection on these genes allowing multiple types (or alleles) to co-exist.

Ghoul and co-workers illustrates the effect of cheating on the (relative) fitness, and how cheating/deception may evolve. "A key factor is whether the fitness of cheats is frequency dependent, such that the relative fitness of cheats decreases as they become more common in the system (*refs*)." ([98] *page 326*). For instance, when cheaters have less relative fitness one may expect equilibrium, and oppositely when cheaters have a higher fitness than co-operators – because there is a cost to cooperation – one would expect the cheating trait to go to fixation.

As there is competition between groups and hence a fight for survival, the amount of cheating matters; cheaters destabilise cooperation ([100]). For instance, groups comprising only altruists will grow faster than groups with only cheaters / defectors, while in mixed groups defectors may have higher reproductive success ([67]); and only when kin selection is weak, can cheaters prevail ([70]). However, this picture seems to be a lot more complicated, as seen from the introduction in Velicer's review: "Selfish social strategies are not limited to mammals with complex behavioural plasticity, … . Cheating is also common in social insects and in microbes with relatively hard-wired social traits." ([100] *page R173*). There are cheating both at the kin and the group levels of cooperation.

Since the presence of cheats imposes a fitness cost on cooperators, mutations that confer resistance to such cheating will be favoured at selection. Cooperators could be selected to reduce the cooperative behaviour that is exploited, even to the point that the cooperative ability is lost. Or alternatively selection can change the cooperative mechanism, for instance, by evolving a receptor molecule that is harder for cheats to exploit; examples are mentioned in [98]. However, if co-operators evolve to become harder to exploit, so will the cheaters, "Consequently, not only does the presence of cheats impose a selection pressure on co-operators, but this can lead to a coevolutionary arms race between co-operators and cheats." ([98] *page 327*); one of the authors' example in this respect is the "brood parasitic cuckoos and their hosts, where the hosts are selected to reject cuckoo eggs, and the cuckoos are selected to circumvent this." ([98] *page 327*).

In his review, Velicer summarises several hypotheses from the literature regarding mechanisms for handling the detrimental effects of cheaters, for instance, that cheating might be restrained at the genetic level within potential cheats: "The most direct way to accomplish this would be to make mutations that cause defection from cooperation intrinsically harmful to fitness ('intrinsic defector inferiority') …" ([100] *page R173*). Velicer then in detail references a study observing such defector inferiority in a cooperative slime mold. It seems that cheaters may be obtained by a single mutation or a few mutations, and defector inferiority has been observed in a number of microbial defectors. One advantage of this mechanism for a social organism is that social interactions or selection against the cheaters are less relevant (although not superfluous), and therefore that efforts may be focussed on other and more productive activities to optimise fitness of that society, such as finding the optimal nest or foraging (in more advances species), etc.

Cheating/deception may be based on aposematism; see this in Section 5.2.5.

### *6.3 Group Dynamics (Acquired Team behaviour), 'Extended Evolutionary Synthesis'*

It took the author some time to realise the topic of this 'cell' in Table 1, while knowing that it had to be about group dynamics and at the same time have a significant aspect of adaptation / learning in order to align with the rest of this perspective in a gradual, progressive manner: What is it that makes a group a team rather than being a collection of mere individuals?

A 'team' is a group of individuals cooperating to achieve a shared goal. Examples are: 1) lions hunting as a group to lay down prey; 2) two birds cooperating to build a nest for their progeny; 3) last year's progeny helping their parents to feed and raise the next generation of their siblings; 4) the guard in a group – this is not a specialised role like in Section 8.2.2 and 8.3.2, since it is accomplished in a round-rabbit fashion amongst the grown-up individuals; and 5) defence mechanisms, such as a herd encircling the new-born calves to defend them from predators.

#### *6.3.1 Acquired Team Behaviour*

The suggestion for topics to be included here are those aspects that are based on acquired team behaviours – that is, learning. That which is learned is not stored in the DNA sequence, and therefore, it can be defended in this place irrespective of the mechanism for learning. Learning is a huge topic in itself and with diverging theories and no definitive and conclusive full explanation. Comparing memory and plasticity reveal that both have short-term adaptation that is converted into a long-term memory by some mechanism (fixation for properties in the plastic trait, and long-term memory, resp.). Of the examples mentioned above, the lions' hunting skills are clearly an acquired skill that is transferred from parent generation to progeny.

Webster and Ward in [66] review the social influences upon individual behaviour, in terms of two behavioural traits, 'conformity' and 'facilitation'. They define 'conformity' as "…the positively frequency dependent tendency of individuals to adopt the behaviour of the majority of their group mates, or their near neighbours within the group, such that they become disproportionately more likely to perform a behaviour as the proportion of others performing it increases (ref)." ([66] *page 761*). The authors say that predators exert substantial pressure, contributing to the selection pressure, in that they disproportionally target oddly behaving individuals (e.g. because pattern recognition will make these particularly visible in the group), and that conformity may operate through simple local rules of for instance alignment with neighbours while attending closely to their behaviours, thus establishing uniformity amongst group members. This could be valid also for both instinctual decision-making and informed decision-making.

Correspondingly, Webster and Ward define 'facilitation' as "Social facilitation occurs when the presence of group mates affects the behaviour of an individual, allowing or causing them to engage in certain behaviours at a different rate, or to perform behaviours that they would not perform at all if they were alone (*ref*)." ([66] *page 762*). The mechanisms driving such facilitation may include reduced perception of risk, decreased investment in vigilance and/or increased levels of competition. The authors reference studies investigating the variation for a given personality trait between asocial and social contexts; they conclude that these studies demonstrate a complex and context-dependent influence of sociality upon individual behaviour, and that "Many behavioural responses are known to be strongly influenced by the number of group mates present (*ref*)." ([66] *page 765*). Further, they find that "In particular, the composition of personality types within the group can feed back to affect both the behavioural responses of its constituent individuals and the way in which the group as a whole functions in relation to the environment." ([66] *page 766*).

Webster and Ward show that there are a number of adaptive aspects in group behaviour ([66]). They say "The genetic and endocrine systems underpinning behavioural responses are highly complex, and the behaviour expressed by the individual, and its underlying genetic basis, may be separated by many hierarchically arranged steps or stages, which themselves may be subject to feedback and interactions with other systems." ([66] *page 760*). They continue by stating that "This complexity potentially allows for a degree of behavioural flexibility, and the environment, including the social context, that an animal experiences can have considerable influence upon its neuroendocrinology and subsequent behaviour (*refs*).".

# 7 Prioritization: Maximising Effectiveness of Evolution

**Abstract.** Function 4, prioritization (see Table 2 and Table 3 in Chapter 14), is concerned with a goal-orientation, thereby putting the emphasis on 'effectiveness'. 'Prioritization' entails selecting between a set of available options, and therefore explicitly involves decision-making processes. The principle is 'survival of the fittest (combination of) decision-making strategies', and the mechanism is 'natural selection of decision-making preferences, explained for the three perspectives: Standard Evolutionary Theory, Inevitable Evolution Theory and Extended Evolutionary Synthesis.

**Keywords:** Natural selection, survival of the fittest, Standard Evolutionary Theory, Extended Evolutionary Synthesis, Inevitable Evolution Theory, collective decision making, decision-making strategies, cultural selection, prioritization, flock behaviour, sheltering, dispersal, migration, fission-fusion societies, speed-accuracy trade-off, quorum sensing

Decision-making processes within a system constitute the means to ensure effectiveness through avoiding appropriate resources. 'Cycling' here is not to be understood in the sense of a wheel going round and round, but of repetition or returning to a point of departure, from which to start again, and again, ..., i.e. recycling. Viewed from a single resource, it is the processing by the seven functions that are operating in a continuous cycle; an analogue in biochemistry is Krebs' cycle, and an example from everyday life is a crossing with a traffic light. The processing at each cycle of the slightly changed resources from previous cycles (see the spiral of progression in **Figure 1**) leads to incremental changes. Such incremental nature in terms of repeated (small) steps is a built-in characteristic at evolution in the sense that we know this.

Effectiveness is concerned with the capability and capacity for bringing about the result intended for the system – that is, securing the system's raison d'être. Thus, effectiveness implies the employment of appropriate decision-making strategies incrementally to achieve a goal. And doing the right thing in any situation – and hence also in an evolutionary context – is closely linked with the maintenance (or optimisation) of fitness, as the opposite – doing the wrong thing – may put fitness at risk, thereby affecting evolution of the individual as well as the group. In some decision-making scenarios, the group decision may be critical for its survival. Examples are the choice of an appropriate nest site, foraging, the behavioural reaction to the appearance of predators, migration dynamics, and the decision to cheat in an otherwise cooperative scenario. Therefore, decision-making strategies cannot/should not be ignored or underestimated as an evolutionary factor in itself, operating indirectly through the endeavour to optimise the capability and capacity for bringing about the result intended for the system – survival and reproduction; and in an evolutionary context the system's fitness is the capability strived for.

**Principle**: Survival of the fittest (combination of) decision-making strategies
**Mechanism**: Natural selection[10] of decision-making preferences
**Emergent property**: Selected phenotypes

In Chapter 3, a couple of examples of individual decision-making were given for the three perspectives: Individuals of the human species may in different contexts show instinctual decision-making (e.g. in caring for progeny), rule-based decision-making (e.g. at craftsmanship) and value-based decision-making strategies (e.g. in culture). The present chapter is dealing with the corresponding collective decision-making strategies.

As expressed by Cronin, "Group-living organisms in a wide range of taxa must make behavioural decisions that affect the entire group while maintaining group cohesion. This is often achieved via a decentralised process known as a consensus decision-making, in which a group response emerges as the product of the actions of multiple individuals …" ([101] *page 1262*).

Given that culture – in short – is 'acquired preferences at decision making'[11], this has in social systems a tight link with the concept of 'effectiveness'. As effectiveness is related to 'doing the right things' (actually, the perceived right things) decision-making preferences determine the individual's actual decisions. Thus, it seems that this function with its focus on decision-making strategies is also concerned with aspects of cultural selection. This is discussed in much more detail in Section 7.3.

This point at a distribution of decision-making strategies over the three perspectives shown in the table:

| **Inevitable Evolution Theory (IET)** | **Standard Evolutionary Theory (SET)** | **Extended Evolutionary Synthesis (EES)** |
|---|---|---|
| Constitutive (instinctual/reflex-based) collective decision making – without learning:<br>• Collective movement patterns, such as:<br>o Shoal/flock behaviour<br>o Sheltering<br>o Dispersal/migration | Collective cognition / acquired rule-based decision-making preferences:<br>• Collective action patterns, such as:<br>o Dispersal/migration<br>o Fission-fusion societies<br>o Misc. decision-making aspects | Collective cognition, acquired value-based preferences in decision-making strategies:<br>• Culture, including<br>o Religion<br>o Institutions<br>o Politics<br>• Informed (adaptive) dispersal / migration |

Definitions in this respect are nicely provided by Ross-Gillespie and Kümmerli in [61]: 'collective decision making' is defined (in its broadest sense) as "… the process by which a group of individuals uses social information to arrive at a state of adaptive group-level coordination. By "social information" is meant signals and/or cues generated by other individuals (*refs*), which could be transmitted directly, or indeed indirectly … ." ([61] *page 1*). Moreover, "By "group-level coordination" we mean anything other than a random distribution of individuals – or their behaviors – in space

[10] Note the definition of 'natural selection' in Section 1.2.

[11] See details in Section 7.3.

or time." ([61] page 1). Jeanson and co-workers present examples of collective decision-making scenarios are selection of nest site, colony emigration, foraging, colony defence and division of labour ([102]).

Note that for all three perspectives, it is the individual that performs the decision-making whether within a group context or not and that some of the decision-making processes literally are group processes, for instance through various positive and negative feedback mechanisms. The text in the following three sections brings more examples.

### *7.1 Collective Decision-Making, 'Inevitable Evolution Theory'*

The basic mechanism in this perspective of evolution comprises instinctual / reflex-based decision-making strategies or decision-making as we for instance see in shoals of fish responding to a predator's attack, and in flocks of birds migrating. Here, it is the synchronous and aligned response by all individuals that constitutes the collective decision-making behaviour in this perspective. As Jeanson and co-workers express it "collective decisions can emerge from the combined action of individuals and the direct or indirect interactions between individuals." ([102] *page 1*), or as Couzin says, "... collective behaviour can arise from repeated and local interactions and need not be explicitly coded as a global blueprint or template [*refs*]." ([103] *page 36*). This is how the present perspective in collective decision making is perceived, while the distinctive characteristic is the absence of cognition within the decision-making process.

Ross-Gillespie and Kümmerli compare microbial decision-making with the collective decision-making of higher taxa. They "conclude that collective decision making in microbes shares many features with collective decision making in higher taxa..." ([61] *page 1*). They say that the individual microbe chemically monitors their close environment and respond according to simple innate (instinctive, inherent) rules. This might seem obvious since microbes have no cognitive ability and have to act basically through a metabolic response to received physical and chemical signals. Unlike most higher taxa, microbes have limited ability for fast, intentional movement (migration) and therefore a favoured reproduction locally leads to large populations of individuals. Since such populations are fairly homogenous, their (metabolic) response to environmental cues can mimic self-organised patterns of decision making at a group level.

Ross-Gillespie and Kümmerli describe three rules that have been explored by simulation; these three rules are related to concentric zones relative to the localisation of a given individual, which guide that individual's decision-making ([61]): 1) if a neighbour entered their outer zone ('zone of attraction') they would move toward it; 2) if such neighbour entered the second zone ('zone of orientation') they would align to its orientation; and 3) if it entered their innermost zone ('zone of repulsion') they would direct them to move away. The authors continue by saying that "Simple sets of rules like these are thought to underlie the complex collective movement patterns observed in fish shoals (*ref*), bird flocks (*ref*) and insects (*ref*)." ([61] *page 3*). They reference a study by Shklarsh and co-workers from 2011 providing hints that similar rules may model collective swarming mobility in a population of bacteria.

Ross-Gillespie and Kümmerli in [61] state that:

1) The flexibility (speed-accuracy trade-off) in microbial response to environmental changes, and which is interpreted as a kind of bet-hedging, may arise

"... because individuals stochastically switch between alternating distinct responses (*ref*)." ([61] *pages 3-4*;

2) In higher taxa, "positive and negative feedbacks work together to modulate the trade-off between the speed and accuracy of the decision-making process." ([61] *page 4*), where the combined action of the two types of feedback generally counterbalance each other to stabilise emerging collective patterns ([102]). Both types of feedbacks are features that have been demonstrated in bacterial decision making, and for instance the positive feedback facilitates the formation of fruiting bodies in amoebae;
3) For quorums and quorum sensing, when the chemical stimulus from the environment (with sensing through receptors) is above a threshold level then the standard response is to induce the production of more of the same signal component – corresponding to positive feedback – and once the receptors are sufficiently stimulated then a 'coordinated' group response in terms of a shift in gene expression is induced;
4) Conflicts of interest (e.g. cheating in the case of aggregation among amoebae to develop a fruiting body): experimental studies confirm that relatedness is a key factor in microbial collective action, which is understandable because the more identical the population of individuals is the better their response to a quorum will be coordinated to achieve a shared goal; on the other side, the differences, and hence the conflicts of interest, will allow for evolution in patterns of fitness;
5) The necessity of aggregation of information for centralised decision making also has an analogue in microbial systems. For instance, the formation of fruiting bodies in starving amoebae, where – at the aggregation – a positive feedback loop is generated, which establishes a chemical gradient that individual amoebae follow, until a dense aggregation is formed.

Couzin in [103] discusses the scaling from individual to collective behaviour and ways these operate. In microbes, this may – as also found in more advanced species – happen through various feedback mechanisms. For instance, Pseudomonas aeruginosa exhibits both positive and negative (local) feedback mechanisms, implemented through secretion of various chemical components.

Not all decision points require instant action for survival, and consequently, there are other approaches for accomplishing decision making, more like consensus decisions. Consensus decisions frequently employ a 'quorum' mechanism, whereby the likelihood that a given individual will undertake a specific action increases markedly once a threshold number of individuals is already performing that action. Information exchange through the quorum sensing systems "... is overwhelmingly chemical in nature, whereas in more complex metazoan it may be audial, visual or tactile, etc." ([61] *page 8*). In microbes the signalling is effected through receptors and with subsequent metabolic responses to achieve a certain goal. Ross-Gillespie and Kümmerli concludes from their review of decision-making mechanisms that many features for decision making among microbes are shared with collective decision-making in higher taxa; see these in the next section.

"Ratio dependence has been found in a wide range of taxa, and to apply to a range of perceptions, including tactile stimulus (*ref*), visual quantification of number (*refs*), auditory discrimination (*ref*), task repetition (*ref*), sucrose concentration (*ref*), visual contrast (*ref*), and abstract concepts such as the estimation of price (*ref*).", as expressed by Cronin in [101] (*page 1262*). It has been proposed that spatial, temporal and numerical sensory information share a common mechanism of magnitude estimation ([101]). Ratio dependence also applies to the quantification of group size as an important instrument in the tuning of group size which frequently affects group fitness. It has been reported that the ratio-dependent distance estimation applies to slime mould, and hence that the mechanism is independent of a neural system ([101]). This kind of decision-making process can be adjusted to suit the environmental conditions, and they are in some cases proportional to the group size, for instance as a function of a ratio between stimuli that increases above a threshold to call forth a decision.

Planas-Sitjà and co-workers study the collective decision-making process for the cockroaches (*Periplaneta americana*) in relation to sheltering ([82]). Cockroaches are characterised by a self-organised process of aggregation, rather than individuals leading or following in the process. The authors emphasise that this does not exclude an aggregation pheromone as the underlying mechanism, at which those producing more than others will contribute more than others to the collective decision-making that lead to aggregation. An alternative hypothesis for the aggregation mechanism is variation in the response to the pheromones from individual to individual owing to differences in their threshold response. According to Cote and co-workers, individual differences in exploratory behaviour are found to be consistent. "Artificial selection experiments over four generations produced fast exploring/bold versus slow exploring/shy individuals, thus demonstrating that exploratory behaviour and copying style are heritable (refs)." ([104] *page 4067*).

Planas-Sitjà and co-workers study the group aspect versus the individual behavioural patterns in collective decision-making for their cockroaches ([82]). They definitely see variation in the cockroaches' individual behaviour even if there is a collective personality at the group level. The individuals vary with respect to sheltering (i.e. their tendency to explore the environment). The "…group personality, which arises from the synergy between the distribution of behaviour profiles in the group and social amplifications, affected the sheltering dynamics. However, owing to its robustness, personality did not affect the group probability of reaching a consensus." ([82] *page 1*). The point here and now is that an individual's behaviour is different when he/she/it is alone as compared to its behaviour when it is member of a group. And naturally, this goes both ways, as the group behaviour also depends on the personality of the majority of members or the leaders of a group. Consequently, the decision-making processes of an individual influence group behaviour and thereby also the fitness aspects of both the group and its members as individuals.

Further, the group personality influences the exploitation pattern of the environment; a social group is not the mere sum of individual behaviours, but is modified by amplification. Thus, the sheltering dynamics is sensitive to the composition of the group: groups with homogeneous individuals are more likely to perform particularly fast or slowly in the aggregation processes. Planas-Sitjà and co-workers caution the interpretation while summarising findings in the literature that behaviours, which are more sensitive to the environment, and therefore more flexible, tend to be less repeatable than behaviours under morphological or physiological constraints ([82]). The authors observe that the cockroaches showed behavioural stability over a week and state that they cannot exclude epigenetic factors.

We mentioned in Chapter 5 pheromones as a means for communicative interaction, and this may in a group context serve as a means for achieving and maintaining the group behaviour. Couzin in [103] reviews how ants use pheromone trails to coordinate activities within a group, which in some cases include multiple pheromones, where variation in volatility adds a time-scale to the trail information laid out since the newest information has the strongest impact, and this way the group avoids becoming trapped in suboptimal solutions. The decision mechanism evolved is primitive but efficient: when foraging for food resource, the ant finding the nearest source returns home the fastest; another one follows the trail and return equally fast, so the closest source of food will immediately achieve the most intense pheromone trail.

Pelé and Sueur review experimental findings exploring trade-off mechanisms applied for prioritization between alternative options, similar to the choice scenario of 'one bird in the hand is better than two in the bush' ([105]). They discuss factors like delay/speed in decision making as well as decision accuracy and risk[12]. The mechanism of the 'Diffusion Model' implies that when the relative advantage of a given alternative exceeds a threshold then that decision is chosen. According to Pelé and Sueur ([105]), this model is the only one able to explain the speed-accuracy trade-off (in binary or n-ary decisions). The Diffusion Model seems valid for a large variety of decision situations and for species that range from monkeys over ants and bumblebees to slime mold, both for the individual and the collective decision making. However, the authors also mention that other types of decision making may lead to other mechanisms, involving for instance survival mechanisms; one such could be a heuristic mechanism with simple exclusion of less advantageous alternatives, and specifically for survival this could also be the constituent decision-making mechanism.

In summary, collective decision making within this perspective seems to primarily be performed by means of a quorum sensing in terms of chemical cues, ratio-dependence and/or threshold mechanisms, but also potentially involving bet-hedging, trade-off mechanisms and simple rules. This does not exclude the influence of individuals' behaviour and personality.

### *7.1.1 Decision Making in Relation to Dispersal/Migration*

Dispersal (or migration) is defined as the "active or passive attempt to move from a natal/breeding site to another breeding site." ([106] *page 209*).

Dispersal is divided on three behavioural stages in succession ([104], [106], [107]): a) departure from the current patch, b) movement between patches, and c) settlement in a new patch. However, there is not a single behavioural pattern that covers all species and all cases for the various stages of dispersal, showing that it is a complex issue. What belong here are the mechanisms influencing the decision making in relation to passive dispersal.

In short, the following are all factors at dispersal and may be stable over ontogeny and/or across situations: stochasticity and/or various differences in behavioural types or behavioural syndromes, such as behavioural profile (e.g. boldness, shyness, aggressiveness, foraging, neophobia, and proactive-reactive strategies). Activity patterns as well as some kinds of social behaviour (altruism and cooperative tendency belong in Section 7.2.6, while mating decisions and parental caring have adaptive decision-making

[12] Pelé & Sueur use the concept of 'risk' in the sense of "when actions may lead to different possible outcomes" ([105] *page 546*).

strategies involved and hence belongs in 7.3.2), and risk management (e.g. avoiding predators or competition for resources) belongs here, when this is purely instinctual. Overall, dispersing individuals are not a random subset of the population ([104]).

An example in this respect involving humans is concerned with lactose tolerance. Lactose tolerance has long been considered a culturally-determined trait in humans, but this is now debated, see for instance [108]. Hypotheses originally suggested that lactose tolerance co-evolved alongside the cultural adoption of milk-drinking where lactating domesticated animals are an integrated part of the conditions for living (i.e. belonging under 'culture'), or as adaptation to a new source of food under extreme climate conditions (i.e. belonging under 'plasticity'). However, an alternative hypothesis that is gaining evidence now is that the lactose tolerance is related to patterns of migration carrying the relevant mutation (see e.g. Allentoft et al. [109]) rather than the conversion from hunter-gathering to farming. If the original hypothesis of cultural adaptation as the mechanism is valid, then this example belongs in Section 7.3; if the new hypothesis is valid, then it may belong here under the Inevitable Evolution Theory under dispersal – the distinguishing feature then would be whether the dispersal is based on an active and intentional decision or passive.

## 7.2 *Collective Decision-Making, 'Standard Evolutionary Theory'*

The basic mechanism in this perspective of evolution comprises acquired, rule-based (informed) decision-making strategies, employing cognition.

### 7.2.1 *Scaling from the Individual to a Group*

Couzin in [103] discusses the scaling from individual to collective behaviour in terms of behavioural tendencies within the zones discussed in the previous section: "Near-range repulsion from others enables collision-avoidance and maintains individual personal space ...", and "A relatively long-range attraction maintains group cohesion, minimizing potentially dangerous isolation ..." ([103] *page 37*). Further, Couzin describes that the collective states can be dependent on previous history, "This demonstrates that animal groups can exhibit a form of hysteresis, or 'collective memory'." ([103] *page 37*).

In social insects, according to Jeanson and co-workers the scaling from individual to collective behaviour "... could be explained by self-organized mechanisms based on the use of simple rules by individuals relying solely on local information, and on the direct and indirect interactions among these individuals (*refs*)." ([102] *page 2*). The combination of direct and indirect interactions that are not mutually exclusive can further enhance amplification and thereby enforce collective decision making. Further, "In non-linear systems, fluctuations at the individual level, even small ones, can lead to profound changes at the collective level, highlighting the fact that noise and stochasticity are intrinsic to any collective decision (*ref*)." ([102] *page 6*); and undirected noise can suffice for a system to behave adaptively to changing environments by facilitating transitions. Still, the combination of "...simple behavioural rules are able to account for the production of complex collective patterns." ([102] *page 11*).

### 7.2.2 *Overall Mechanisms of Animal Decision-Making*

Couzin in [103] reviews the mechanisms of collective cognition in relation to decision making in groups of animals, and step by step he draws an analogy of these with specific neuronal processing mechanisms in the brain, although the underlying detailed mechanisms are not known. He says that: a) multiple stable modes of collective behaviour can co-exist for given individuals; b) alignment among individuals (a tendency to move in the same direction as near-neighbours) can enable coordination of change in direction, and such amplification of local fluctuations through positive feed-back is important in case of threats; c) incorporating negative feedback can prevent over-sensitivity ('informational cascades') of collective responses to individual error or environmental noise and can enable distant patterns to be observable in relief of undesirable local fluctuations; d) quorum mechanisms, threshold mechanisms and aver-aging are employed to improve the accuracy of individual decision-making by integrat-ing own preferences with that of others.

Similarly, in case of searching for a new nest, scouts will make the search and subsequently recruit fellows with an eager that depends of the site quality and may culminate in carrying fellows to the site to show commitment; thus, a graded signal where the process speed is tuned according to urgency ([103]). Social insects like ants and honey bees employ quorum sensing in connection with the collective decision-making process on deciding among potential nest sites ([103]). For instance, honey bees will initiate swarming once 10-15 scouts are in a single location: when a colony needs a new home scouts go out to find a suitable new location, and the voting is achieved by recruitment of nest mates to each their favourite site, and once a threshold number of scouts is reached the quorum is achieved on that nest site. Cronin in [101] reports from his investigation with ant quorum sensing that a clear positive association between quorum threshold and colony size exists without an observable influence of experience. Thus, ants are employing an analogue to the chemical trace with a magni-tude mechanism involving the ants' ability to discriminate quantity.

Conradt in [110] reviews models for studying animal collective decision-making with a focus on information uncertainty and conflicting preferences as factors in the decision making. The reason that this is important in an evolutionary context is that the difference between a wrong and a right decision may be the survival of the individual and/or the group. It is obvious from the review that there is still a lot of work to do in this area, and in particular, that the combination of the two factors (information uncer-tainty and conflicting preferences) is unexplored within the literature ([110]). Conradt discusses models dealing with uncertainty within the decision-making information, for instance:

1) Quorum response mechanisms: "...the probability of an animal to choose a particular option increases steeply once a threshold 'quorum' of other animals has chosen that option." ([110] *page 227*). However, even if simple animals can, and do, use quorum responses the behind cognitive mechanism(s) are not clear. It seems that beyond simple all-or-none approaches it may involve even complicated ratio estimates (see this concept in the previous section). In very large groups the quorum model may become very slow and thereby inefficient for certain decision-making situations, such as the attack by predators;
2) Conradt says "...it is generally accepted that movement decisions in large groups are based on local self-organizing interactions between neighbouring individuals that result in global cohesive and synchronized group movements [*refs*]." ([110] *page 228*), while the behind cognitive mechanism as well as the penetration of the outcome is still unclear;

3) The independence-interdependence model is addressing noisy but independent information combined with interdependencies between the decision makers; and because of the interdependency information pooling may be facilitated. A study that Conradt references concludes that "...without interdependence, the rapid convergence to a consensus would be undermined... , without independence, a consensus would still emerge, but it would no longer robustly be in favour of the highest quality option. ... only when independence and interdependence are combined in the right manner can animals achieve high collective accuracy." ([110] *page 230*).

It appears that the quorum mechanisms and the self-organising group movements exist in a form that has no cognitive element within the underlying mechanism (i.e. belonging in Section 7.1) and another version with a cognitive element.

Petit and Bon in [111] review decision-making processes in relation to collective movement. They define 'collective movement' as "... a group of animals that decide to depart/move quite synchronously, move together in the same direction (...) and maintain cohesion until the group stops moving or starts a new activity, all resulting in a change of location." ([111] *page 635*). They summarise the literature, deducing that "Collective patterns or properties at the group level arise as a consequence of local interactions between individuals, without centralised control, a common pre-determined goal to be reached or reference to a global pattern at the individual level." ([111] *pages 635-6*). Such behaviours (species-specific) that lead to initiation or maintenance of collective movements may be of a mimetic kind or follow a specific body-language (verbal and/or physical), so it seems to align with Conradt's described models. Petit and Bon express that recent experiments show that interactions among small sets of individuals (even in species that dwell in large colonies) exhibit the basic ingredient of collective activities, and that such collective behaviours result from a multitude of local interactions between individuals – that is, from observations of other members' activities at a short range. It turns out that decision-making among many individuals involved are slower, but tend to be the most accurate and adaptive because they efficiently utilise the diverse information possessed by group members.

### *7.2.3 Handling Conflicting Information*

In addition to the uncertainty factor, the review of Conradt discusses models addressing conflicting preferences, where the issue in general is about 'who decides' and about the timing of a decision, for instance:

- The 'leader-follower model' and the 'pair-synchronization model': both models look not only at the group perspective, but also at that of the individual: conflicts of interests between individuals may arise for instance when the timing of various activities (such as the desire toward foraging and resting) differs between given individuals. In both cases, the strength of a need may urge an individual to take a transient leadership, where the latter model operates with a time window around an optimal point in time for a given activity and which may enable flexibility in the synchronisation of activities and thereby reduce the conflict. In groups of up to ten members it was found that stable dimorphisms of leaders and followers evolved, and further, "The proportion of leaders in the population increased with the degree of conflict, while the degree of coordination decreased with the degree of conflict." ([110] *page 235*).

- The 'group-level model': Both shared decision-making, or shared leadership, exists in nature (e.g. for birds, primates, bats, carnivores and ungulates), and dictatorial decision-making exists in nature (e.g. dolphins, elephants, primates and birds). The group-level model suggests that "... for most biologically relevant assumptions about the shape of the fitness cost function, the expected net costs are lower to a group that makes shared decisions than to a group with a dictator." ([110] *page 230*).
- Self-organized system models of collective decision in large groups with local interactions; the 'Leading-according-to-need' model: Like for the Leader-follower model, "Even in large groups, individual animals may increase their influence on collective decisions by strategically changing simple behavioural parameters, in particular their assertiveness, movement speed and social attraction range." ([110] *page 235*). Moreover, the author states that it is "...likely that group movements are led by those animals for which reaching a particular destination is either most crucial or group cohesion is least important." ([110] *page 235*). Consequently, such decisions may be effected in the absence of knowledge of or the communication about the needs of other group members, and even without the assumption of altruistic behaviour.

### *7.2.4 Moderating Mechanisms for Decision-Making*

Ross-Gillespie and Kümmerli (with [61]) supplements Conradt's view by taking a different approach to their review, while discussing microbial decision-making in the perspective of the collective decision-making of higher taxa (see Section 7.1), focussing a little more on the mechanisms behind the decision-making:

- Speed-accuracy trade-off: maximal accuracy in decision making is not always feasible because of urgency combined with the need for information gathering to be able to assess alternative options for the decision making. In a group context there is additionally the issue of mechanisms behind distributed versus centralised decision-making, i.e. the 'who decides' as discussed above for Conradt. Ross-Gillespie and Kümmerli suggest that "The best adapted decision-making processes should therefore allow for some degree of bet-hedging, and also be adaptable and reversible (*ref*)." ([61] *page 3*), and that under stress varying over time and space, both individuals and groups will be under selection to be "flexible in their collective decision-making, having the ability to transition quickly between decision modes favoring high accuracy and those favouring high speed." ([61] *page 3*).
- "...positive and negative feedbacks work together to modulate the trade-off between the speed and accuracy of the decision-making process." ([61] *page 4*): Positive feedback enables a direct amplification of an information carrying signal and thus may speed up the decision-making process, while negative feedback hamper the accumulation of information and ultimately favours a more accurate response. Both kinds of feedback are well-known in social insects and higher taxa and are an integrated part of the communication

scheme; the example of bees searching for a new nest site has already been mentioned.

- Quorums and quorum sensing (see the discussion on terminology in Section 5.1.1.1): there is a barrier mechanism enabling coordination at a group level in that a group response is achieved and only achieved in case a sufficient number (or fraction) of group members react to the quorum signal – democracy is one such example.
- Aligned versus conflicting interests in the decision-making: conflicts of interests in general will destabilise a cooperative system because conflicts block the decision making; the blockage depends on the magnitude of the conflict and hence is a function of the balance between an individual's advantages from selfish independent action (e.g. cheating) as opposed to a cooperative group action. Mechanisms to avoid conflicts and maintain collective actions are, for instance: a) increasing relatedness enhance the incentive toward altruism; b) introducing a cost on cheating, e.g. by repressing competition ("For essential traits, negative-frequency-dependent selection intrinsically limits the spread of cheats, because at high cheat loads collective actions can collapse – to the detriment of all." ([61] *page 7*)); and c) 'incentivising cooperation' through benefits associated with the cooperative action; an example is the migration of birds, where the formation of the group induces an increase in energy efficiency from slip-streaming and hence reduces the risk of exhaustion at migration for all individuals in the group.
- The role of leaders/followers and the necessity of aggregation of information, from local to global, to enable the centralised decision making; this topic is clearer dealt with in Conradt's review, and hence see above bullet lists. Aplin and co-workers add in this respect that leadership tendencies within individuals seem to be consistent and repeatable among individuals with a proactive behaviour, so this is a behavioural trait of significant relevance for the collective decision-making ([112]).

### 7.2.5 *Personality and Group Decision-Making*

Aplin and co-workers with their simulation model suggest that "... simple interaction rules are sufficient to explain the patterns of collective behaviour observed ..." for the wild great tits ([112] *page 4*), also saying that while all individuals exhibited some degree of collective behaviour, this varied with personality (proactive versus reactive tendency), where "reactive individuals have a greater social attraction to conspecifics and are more likely to use social information." ([112] *page 5*). The authors further say that the personality in wild great tits "is thought to relate to a differential response to risk-taking, with proactive (FE) birds engaging in potentially highly rewarding behaviour with an higher associated risk, and more reactive (SE) individuals favouring a lower productivity by low-risk strategy." ([112] *pages 6-7*). The authors conclude that there is "...an increasing body of evidence that social behaviour and collective decision-making may not just reflect immediate costs and benefits, but may also be an outcome of intrinsic behavioural differences between individuals ..." ([112] *page 7*).

One aspect of collective decision-making according to Webster and Ward may be 'leadership', in that a bold personality trait may consistently lead an individual to acting as leader ([66]). Still, this is not leadership in the human social sense, which is relatively scarce in nature, but may for instance be temporary leadership (initiators), leadership distributed on many individuals and/or based on a kind of voting process; see also [111], who also mentions leadership based on experience. Another aspect is the 'social organisation', also involving self-organising processes, but this aspect is not yet well-explored. A third aspect of the collective decision-making is 'producer-scrounger interactions', which is concerned with the acquisition of information from social sources, by monitoring or interacting with kin and/or via exploration of the environment, each followed by a consequential act that may have a range of costs and benefits. So, indeed personality may play a significant role in determining how individuals (of some species) act based on social information, and which therefore has significant implications for the individual fitness as well as group fitness and consequently also group dynamics.

### *7.2.6 Decision Making in Relation to Dispersal/Migration*

"Dispersal is a strategy to increase fitness in a heterogeneous landscape by changing the environment in which an organism lives." ([107] *page 218*), where dispersal may have a stabilising effect on unstable dynamics.

"Through simply moving from one habitat patch to another, the dispersal of an individual has consequences not only for individual fitness, but also for population dynamics and genetics, and species' distributions (*refs*)." ([107] *page 206*). The authors say that "Kin selection may favour dispersal as a mechanism to reduce competition between kin at the natal site." ([107] *page 208*); the issue (within the models studied) being that the parent generation dies after reproduction and then the offspring has to compete for the now vacant patch. Further, Bowler and Benton state that condition-dependent dispersal often is superior to unconditional, fixed strategies, of which the latter type belongs and is discussed in Section 6.1.1, while the former requires a cognitive process and hence belongs here or in the similar section in 7.3.2, depending on the nature of the underlying process.

Bowler and Benton in [107] review existing knowledge – at that time – about dispersal, both from experimental and theoretical studies. A number of factors causing and influencing dispersal have been identified: a) ultimate causes, such as patch population density and demographic stochasticity influencing the kin interactions; inbreeding avoidance, and habitat variability (including its intrinsic quality); and b) proximate causes for i) emigration, such as density, food availability, interspecific interactions (e.g. with parasites and predators), sex ratio, patch size and isolation, relatedness, and matrix habitat (where the cost of dispersal may be assessed prior to emigration); for ii) the inter-patch movement (movement between patches), such as matrix habitat, search strategies, habitat cues; and for iii) immigration, such as patch size, isolation, habitat cues.

Non-random systematic search strategies have been documented in the meadow brown butterfly and the gatekeeper butterfly; see [107]. These butterflies – when released in a new habitat far away from their regular habitat – literally fly in a petal-like loop from start and back to the starting point. This allows then to scan the area for a suitable patch and return to the starting point if not. "Modelling work has revealed these non-random search strategies often to achieve a greater dispersal success than random strategies (*refs*)." ([107] *page 213*).

Bowler and Benton in [107] mention some additional propensity influencing dispersal, such as sex, developmental stage and body size or condition. For instance, in some species it is the female that migrate, in others it is the male that migrates. Also, if dispersal requires bodily reserves, then it seems to be the larger individuals showing a tendency to migrate, for instance, because the reserves of body fat can improve the chances of surviving the dispersal processes as has been shown for the naked mole-rat.

### 7.2.7 Decision-Making in Fission-Fusion Societies

The mini-review of Kerth, [113], discusses fission-fusion societies, where subgroups are formed dynamically depending on individual preferences. He concludes that "By forming temporary subgroups that represent individual preferences better than a group as a whole fission-fusion societies avoid a permanent break even in situations where conflicts among their members are to strong to reach a consensus." ([113] *page 663*).

Kerth says that "…group decision-making processes in fission-fusion societies are not fundamentally different from those in cohesive animal societies that depend more strongly on reaching a consensus." ([113] *page 663*). The Bechstein's bats belong to the category of fission-fusion societies in that a colony dynamically forms subgroups of individuals on a temporary basis and which regularly split to form new groupings. Also bison belong in fission-fusion societies ([114]). Merkle and co-workers suggest three mechanisms that can explain how fission-fusion dynamics shape the dynamics of space usage ([114]): 1) individuals consistently learn about the quality of new sites by joining other groups; 2) such transfer of beneficial information will increase favourable decisions, as there is a tendency of the individual to follow those that are known to or seem to know the good sites, thereby increasing the fitness by reducing the uncertainty of the food sources; and consequently, 3) knowledge transfer of the best sites with subsequent adjustment of individual behaviour will generate resource competition. New individuals in a group may acquire information about the quality of patches in the area by observing the behaviour of others, and this way such individuals "…can become leaders because of their dominance status or because conspecifics recognise their knowledge…" ([114] *page 804*).

Fleischmann and co-workers experimentally explore the decision-making behaviour of female Bechstein's bats to observe the influence of conflicting information on the groups' choice of daytime communal roosts, which for that species is selected on a daily basis (i.e. a 'fission-fusion society'); see [115]. Various proportions of the group members were exposed to one of three unpleasant stimuli when investigating a candidate place. The authors show that the rules applied to make group decisions vary with the level of conflicting information/personal preferences among the group members. The study also reveals that the bats consider not only their individual preferences, but also the preferences of other colony members when making group decisions, except for situations with a very strong conflict where the individuals would follow own preferences rather than seeking a compromise and thus impeding coordinated actions.

### 7.2.8 In Summary

There is a wealth of mechanisms involved in collective decision-making, and which may not all be mutually exclusive. Moreover, there seem to be a series of mutually orthogonal factors influencing such decision-making: personality traits, type of group / society, and mechanisms for handling conflicting preferences.

### *7.3 Cultural factors in Evolution and Survivability, 'Extended Evolutionary Synthesis'*

Darwin talks about "… under changed *habits* or conditions of life…" as factors in the natural selection, ([20] *page 567*; the Italics was added for emphasis) and argues that his theory of natural selection according to fitness still holds in that context. Darwin also mentions psychological factors, but does not in this work explicitly include or exclude the social/cultural context of an individual as a factor in the selection process.

The basic mechanism in this perspective of evolution comprises value-based decision-making strategies – that is, as these express themselves in cultural behaviour. It is the element of learning that suggests culture's place here.

'Culture' is concerned with our up-bringing of our children and the imprinting of their decision-making preferences, consciously and unconsciously, by means of a learning mechanism that is facilitated by overlapping generations:

> "Our understanding of the concept of 'culture' may be expressed shortly this way: "By cultural behavior, we mean the stability across generations of behavioral patterns acquired through social communication within a group, and valued by the group" (*ref*, cited and discussed in *ref*). Culture is the style of working in the field or the mental, tacit (learned) behavioral pattern behind the style of working (*refs*). Thus, culture is guiding the preferences; culture is what comes before starting a discussion of strategy, and so forth, in a chain of causal events toward problem solving. When specifically talking of the interpretation of culture in an organizational context it means "the acquired preferences in problem solving", where problem solving should be understood in the broadest sense and not only as problem solving in a profession oriented perspective."
>
> (citation from ([15] *pages 289-290*).

The capacity and capability for problem solving affects fitness directly in various scenarios. Moreover, a system comprising a set of cultural preferences is evolved to provide optimal solutions overall to the local conditions and traditions. Whether or not other species than humans show signs of culture is unknown to the author.

A couple of substantial reviews (see [73], [116]) address the controversial topic of culture-based evolution. The definition of culture that Smaldino uses is slightly different from ours, but not deviating enough to have an impact on the discussion in this section: "… information capable of affecting *individuals'* behaviour that they acquire from other members of their species through teaching, imitation, and other forms of social transmission" ([73] *page 243*). Smaldino's emphasis is on the evolutionary competition between cultural groups. Richerson and co-workers similarly address the issue of selection on cultural groups, expressing that human cultural groups have all the key attributes of a Darwinian evolutionary system ([116]).

Here, one needs to emphasise that culture involves not only a single decision-making strategy, but a series of dimensions, each with dedicated decision-making preferences; see for instance [16], [17], [18], [19], [117], as well as a brief overview in [118]. Thus, a single culture comprises a pattern of decision-making preferences that compares to a super-group. As part of this come institutions, religions and profession-oriented cultures, aligned with the national / regional culture in which they are operating.

The set of decision-making preferences is acquired from individual to individual and adjusted (evolving) over generations to make individuals within a (sub-)population optimally fit for their environmental (here: social) conditions. Furthermore, cultural decision-making preferences seriously affect the individuals' behavioural patterns toward kin within the group and thereby may extend/refine/modify the factors within kin selection: The cultural framework of Hampden-Turner and Trompenaars for measuring cultural aspects for instance includes a dimension with the two extreme concepts of ‘communitarian’ as opposed to ‘individual’ (see [16], [17], [18]). These authors provide convincing measurements in this respect. The communitarian attitude (the collective ‘group first mentality’, where serving group benefits is preferred over serving one’s own personal goals, corresponding to an enforced altruistic behaviour) characterises for instance Japan. On the other extreme is the individualistic attitude, which characterises the USA: “Americans believe you should “make up your own mind” and “do your own thing” rather than allow yourself to be influenced too much by other people and the external flow of events. Taken together, these are the prime attributes of entrepreneurship: the self-determined individual tenaciously pursuing a personal dream.” ([16] *page 48*). That is, the cultural emphasis on altruistic behaviour is loosened in an individualistic culture, encouraging individual freedom and responsibility. This no doubt influences the tendency toward altruism in a negative direction, but of course on the other hand has the advantage of entrepreneurship. Similarly, the most communitarian of their investigated cultures (Japan) strongly favours altruistic behaviour.

Here some may argue that why does so drastic cultural differences exist when some apparently are better than others; the answer is that it is not a single cultural characteristic of a group that determines the fitness, but the combination of cultural traits. Hampden-Turner and Trompenaars’ framework emphasises seven distinct dimensions, while other researchers suggest a different set and a different number of cultural dimensions; see for instance [19] and [117].

Now, the question is whether there are links, strong and/or weak, between culture and gene-based evolution, or whether culture constitutes a separate evolutionary line with its own selection mechanisms? To answer the question, it is necessary for a minute to go broader than selection for the individual.

Richerson and co-workers in their review conclude that “The current direct evidence for culture-led gene-culture coevolution is only strong for genes that do not directly affect behaviour …” ([116] *page 16*), while expressing that studies on culture-based social selection on human genes are only in their infancy. Further, “Evidence currently exists that supports the hypothesis of culture led gene-culture coevolution for a few simple genetic traits whose function is well known, but unfortunately not yet for genes related to behaviour.” ([116] *page 6*). The authors propose that cultural group selection can exert selection on genes via culturally transmitted cost and benefit schedules (coevolutionary social selection). Moreover, they dare propose that culture led gene-culture coevolution could produce much the same result as group selection more directly on genes. They reference a recent study of Perreault demonstrating that cultural evolution evolves considerably faster than genetic evolution, and therefore, culture will be the leading process in gene-culture co-evolution.

Richerson and co-workers in [116] emphasise a comparative study between chimpanzees and humans, showing that the former only imitate (and hence learn) behaviours necessary to reach a goal, while children of humans meticulously imitate all of an adult’s behaviours. By such extensive imitation, the transgenerational transmission of information and behaviours for a culture to thrive is present. Further, they conclude “Thus, a scenario by which the capacity for complex trustworthy language evolved by

CGS and culture-led gene-culture coevolution (…) is plausible."[13] ([116] *page 14*), meaning that at least humans have the language capacity to further support the transgenerational transmission of information and behaviours.

Smaldino says "Humans seem built to learn from one another, and most differences between groups of humans appear to be largely the result of learning rather than genotype" ([73] *page 249*). A commentary to Smaldino's paper, [119], mentions the problems in the notion of 'inheritance' within a cultural context, because such cultural inheritance is not genetically based. However, the ability to learn is genetically-based and hence transmissible, and so are the mechanisms behind patterns of decision-making; or at least the hypothesis is that the cultural patterns must have a foundation within the phenotype (i.e. including the (epi-)genetics), otherwise such extremely consistent and richly detailed cultural behaviour would not be feasible. The main point in Davis & Margolis' commentary is the problem that group-level traits are not inheritable in a genetic sense, while the cultural multilevel selection "the cMLS framework can fully explain this in terms of facts about inheritance among individuals."[14] ([119] *page 258*). Still, the indirect effect of decision-making strategies on fitness suggests that culture is a significant factor in evolution; just think of how science has contributed to the competitive fitness of the western cultures.

At least there are parts of human behaviour that are correlated with the genotype or epigenotype, as also discussed earlier. There are examples of events/conditions related to culture that affect human behaviour significantly, for instance (see also the example on Borderline Personality Disorder in a little more detail in Section 8.3.3.1):

1) Smaldino brings an interesting example: "Perception is constrained in part by our biology, but culture also constrains even our basic perceptions of a situation … (*refs*)." ([73] *page 251*). To support this statement, he brings the example, "…(*ref*)… showed American and Japanese university students animated underwater scenes with a focal fish. In a recall task, Americans were much better identifying fish they had seen independent of background information, but Japanese students were much better at remembering details of the background scenes." ([73] *page 251*). Considering that our perception (visual and other) provides the input for our decision-making, such difference is noteworthy. Further, Smaldino explicitly says "Cultural differences in patterns of perception and memory fit larger cultural differences in epistemology and styles of thinking that exist between East and West (*ref*)." ([73] *page 251*). Consequently, such cultural differences will penetrate into the cultural patterns of decision-making preferences, and hence affect evolution.
2) We brought forward another example in [1] (*page 466*) concerned with the effect that a father's obesity has on his daughters via epigenetics (transgenerational epigenetics): "It has been shown that daughters of obese fathers tend to get obese, but more than that, they exhibited impaired glucose intolerance and insulin resistance. The findings of Ng and co-workers that a father's diet can affect daughters' health are extensively discussed in Skinner (*ref*) to find a plausible mechanism. The author concludes that the findings indeed suggest the involvement of an inheritable epigenetic molecular mechanism. ..." (end of citation). Obesity is clearly a problem in the Western cultures and not in the

[13] CGS is an acronym for 'Cultural Group Selection'.

[14] cMLS is an acronym for 'cultural Multi-Level Selection'.

poorer Eastern/developing cultures. So, the culturally determined conditions of obesity will penetrate differently in different cultures at evolution, thereby influencing the future gene pool differently in West compared to that of the developing countries.

"Certainly, the fact that DNA methylation constitutes a source of phenotypic variability, together with recent evidence showing that some epigenetic changes can escape trans-generational erasure mechanisms, suggests that these modifications have the potential to influence micro-evolutionary processes." ([46] *page 3*).

We assume that the culture-led gene-culture coevolution that Richerson and co-workers suggest in [116] is real, and consequently, that there may at least in some cases be a genetic or epigenetic foundation behind cultural factors in evolution and survivability.

Religion belongs under the heading of 'culture'; see [116]. The example of extreme Islam (and similarly with other religions practising or enforcing gender-specific norms for behaviour) seems to be the opposite of altruism, although not its antonym, selfishness, but rather religious beliefs, practiced through the relations between kin. Might it be that there is a class of corresponding antonymous behaviours that from an evolutionary perspective apply mechanisms similar to those of altruism but with an opposite sign? A guess would be that here may be many types of behaviours beyond those that were addressed and which have an effect on evolution, positive or negative, for example selfishness, power & greed, missioning and religious dogma imposed onto fellow kin, as well as co-dependency and contextual adaptability.

Acquired decision-making preferences (i.e. imprinted behaviour) are the foundation for cultural selection ([73], [116]). One example is that at present people in the economically favoured Western countries have better conditions for survival (i.e. better fitness within the local context, affording food, housing, medicine, education, etc.) compared to other cultures, and they are in general able to support and hence pass on their favourable conditions to their off-spring, for instance in terms of a familial/social tradition of seeking an education and/or co-funding such education, and educating boys and girls alike.

Smaldino – in the context of cultural evolution outlines some mechanisms that will maintain and evolve traits ([73]):

- Psychological mechanisms for cognition and perception providing the structural support for social learning
- Transmission isolating mechanisms ensuring that cultural identities remain relatively stable, even when individuals from different cultures interact
- "Humans have species-specific cognitive and perceptual mechanisms that allow for scaffolded social learning, which likely facilitated the emergence of cumulative culture" ([73] *page 250*) – that is, the mechanisms for maintaining a group trait exist

Cultures have an analogue to greenbeard marks, namely symbolic markers, such as dress code, religious marks, dialects and alike ([116]). Two more strong mechanisms are the development of norms and institutions (in the broad sense, thus including religion), where the moralistic norms are used a) "to control antisocial behaviour and thereby dampens within-group phenotypic variation and amplifies variation between groups." ([73] *page 246*); and b) to maintain a cultural identity among members of the group. Once an institution is established, it may enforce cooperation through reciprocity (direct and indirect), reputational exclusion and punishment, acting to induce individual conformity. Further, Smaldino says "Interdependence sustains cooperation and provides a stable environment of mutual aid in which differentiation, division of labor, and complex group organization can emerge." ([73] *page 248*).

The question early in this section was "whether there are links, strong and/or weak, between culture and gene-based evolution, or whether culture constitutes a separate evolutionary line with its own selection mechanisms?". The conclusion here from the discussion of various examples aligns with that of Richerson and co-workers that culture-led gene-culture coevolution is a factor in evolution ([116]).

### *7.3.1 Cultural Group Selection*

In the above, culture is defined in short as acquired preferences at decision-making. Such conception comes from the domain of social science and informatics, and therefore, it is not explicated with respect to whether it operates at the individual and/or the group level.

Smaldino cites Darwin's "Descent of Man" for proposing that "at least in the case of humans, many psychological traits related to empathy and altruism would have been selected against at the individual level and so must have evolved because groups of cooperative individuals would outperform selfish groups." ([73] *page 245*). Further, he states that "… the fact that collaborative behaviours, requiring differentiated and structurally organized roles, have played an essential role in the success of human groups." ([73] *page 246*).

Smaldino emphasises "… that group-level traits, which involve organized collections of differentiated individuals, present a unit of cultural selection that is not encompassed by selection on individuals." ([73] *page 243*). This implies – as he states – that "… selection on group-level traits is qualitatively different from selection on groups as defined by traditional multilevel selection (MLS) theory, which does not account for emergent traits based on group organization." ([73] *pages 243-4*), while also saying that the studying of cultural evolution has suffered from an overemphasis on the experiences and behaviours of the individuals at the expense of complex group organisation and behaviour. The previously mentioned commentary of Davis and Margolis (and other similarly valuable commentaries in the same journal issue) points specifically at the problem in Smaldino's formulation, that it is the group-level traits that constitutes the unit of selection rather than the individuals ([119]). Personally, the present author prefers taking a pluralistic approach again with the example of commutarianism versus individualism mentioned above as the main argument. Thus, the issue of cultural selection is perceived as a both/and scenario, and hence cultural group selection is similar to the group selection (see Chapter 6), but may be more complex because of group internal structures and dynamics with more nuances.

Richerson and co-workers say that cultural group selection can favour motivational systems that alter individual incentives, thereby reducing or eliminating the need for altruism ([116]), and thereby they also emphasise the interplay of selection at the two levels. They further discuss the necessary conditions and conclude that "… CGS occurs wherever significant cultural variation exists between neighbouring and competing groups." ([116] *page 11*).

Smaldino concludes in his review that "Once cooperation between individuals evolves, the stage is set – via persistent association, interdependence, and cultural transmission – for the evolution of nuanced collaboration between individuals with differentiated roles in a meaningful social organization." ([73] *page 248*). Further, Smaldino discusses the characteristics of group-level traits from the niche context of group-level traits in cultural evolution: "... a large part of group success comes ... from the organization of a well-defined collection of individuals all participating in a group-level behaviour ..." ([73] *page 246*), and "The properties that allowed one group to triumph or persist against another in these cases did not belong to each individual group member, but rather emerged from the organized interactions between those individuals." ([73] *page 244*).

Psychological mechanisms likely co-evolved through mutual reinforcement with social structures that promoted coordinated communication and organisation ([73]). Human behaviour can become extremely specialised during development through learning and tracing opportunities, and an enabling means in this respect is among others speech, symbolic reasoning and flexible hands with a fine motor function that may enable the hands to serve as tools or for creative purposes.

Mechanisms that allow not only within-group but also between-group differences and patterns of organisation to be maintained and transmitted include among others: a) cumulative learning/transmission strategies and cultural imprinting like in the example with American and Japanese students (see above), and b) ethnic markers and a number of psychological, linguistic and cultural barriers. Further, Richerson et al. list an explicit set of mechanisms that may maintain intergroup variation ([116] *page 4-5*):

- Accurate, rapid social learning – the human capability for teaching and learning allows the cumulative evolution of complex traits
- Conformist social learning – if people differentially copy locally common behaviours, the homogenising effect of migration will be reduced
- Coordination payoffs – if the payoff to any strategy depends on the local frequency of its use, then even sub-populations in identical environments may reach stable equilibria over time
- Punishment of deviant behaviours – if punishment is sufficiently cheap to punishers and sufficiently costly to the punished, punishment can stabilise any behaviour
- Strong prestige bias or one-to-many transmission – disproportionate learning from selected (particularly successful or charismatic) individuals may enable small cultural populations and drift will diversify populations
- Symbolic markers of group boundaries – neutral makers favour attentiveness to these markers and contributes to group differentiation by reducing the chances that migrants are imitated
- Institutional complexity generates inertia and reduces borrowing – tacit knowledge and non-shared knowledge delimit diffusion of institutions between societies

Cultural group factors include religion and political issues ([73], [116]).

In summary, like the relation between kin selection and group selection it seems there is a similar relation for evolutionary competition between individual culture and group culture, where the group institutes a superior wholeness while selection operates at both the group level and the individual level, and where the group culture involves the groups' internal structure.

### *7.3.2 Decision Making in Relation to Informed (Adaptive) Dispersal/Migration*

Dispersal has several constituent elements as seen from the previous two sections. There is the purely stochastic dynamics as discussed in Section 6.1.1, and the dispersal that has genetic elements in its causal decision-making origin as discussed in Section 7.2.6, and then the adaptive, informed dispersal that is the focus in this section.

The perspectival review of Clobert and co-workers in [106] discusses the processes that "...link condition-dependent dispersal, phenotype-dependent dispersal and habitat choice strategies: 1) the relationship between the cause of departure and the disperser's phenotype; 2) the relationship between the cause of departure and the settlement behaviour and 3) the concept of informed dispersal, where individuals gather and transfer information before and during their movements through the landscape." ([106] *page 197*). Several factors are involved, such as aspects of competition, mate choice and habitat quality. The authors emphasise that the decision to leave a patch and settle in another patch often is both condition-dependent (i.e. individuals rely on a set of external cues) and phenotype-dependent (i.e. the inclination to move is correlated with a suite of phenotypic traits). Regarding (1), the observed phenotype traits that differ between dispersers and residents – and which therefore may be involved in the mechanism causing the dispersal – vary for many species and include for instance hormones and neurotransmitters, various behavioural aspects (e.g. explorative nature, sociality and aggressiveness), morphology (e.g. body size and condition) and life history conditions (see this concept in Section 8.3.3.1) ([106]). The authors state that evidence exists in favour of a genetic correlation between dispersal and life-history, behavioural and morphology traits in ciliates, wing-dimorphic insects and one bird species. Further, they reference a series of examples where environmental factors cause phenotypic differences between dispersers and residents, related to kin competition, intraspecific competition and habitat quality, and hence these environmental factors may induce dispersal types. See more details and arguments in [106].

Regarding (2) above (linking the cause of departure with the settlement behaviour), Clobert and co-workers argue that "the production of plastic dispersal phenotype is in fact a general mechanism linking departure decisions with the future breeding habitat selection strategies of dispensers..." ([106] *page 202*), and that dispensers born in high-quality patches are more likely to settle in good-quality habitats than dispensers from bad-quality patches, while in other examples the familiarity of the new patch to previous experiences may be preferable over the real quality of the patch.

Regarding (3) above (informed dispersal, where individuals gather and transfer information before and during their movements), the information gathering may be of several kinds, such as a) gathering information about conspecifics present in the new patch, for instance on morphology (e.g. nutritional state), behaviour and reproductive success; and b) landscape landmarks or other abiotic cues such as thermally and otherwise suitable habitats.

In summary, Clobert and co-workers with their argumentation and examples make it plausible that phenotypic plasticity is a significant factor in dispersal at all three stages, emigration, movement and immigration.

# 8 Differentiation: Survival of the Fittest Orchestration

**Abstract.** Function 5, orchestration (see Chapter 14), is concerned with among others speciation, specialisation (including eusociality) and life-history theory, risk and quality management; all of this related to a strive toward continuous improvement of the system. The principle is 'survival of the fittest orchestration', and the mechanism is 'natural selection of the fittest orchestration', explained for the three perspectives: Standard Evolutionary Theory, Inevitable Evolution Theory and Extended Evolutionary Synthesis.

**Keywords:** Natural selection, survival of the fittest, Standard Evolutionary Theory, Extended Evolutionary Synthesis, Inevitable Evolution Theory, differentiation, speciation, specialisation, eusociality, life-history theory, interspecies dynamics, interspecies cheating, deception, interspecific killing, risk management, quality, symbiosis, parasitism, evolutionary arrow of time

The Mereon Matrix shows us that an advanced system involves specialised roles; we see these within the matrix in terms of the 7 functions. 'Orchestration' is a term designating all aspects of the ability of an organisation to organise resources and activities appropriately in sub-systems. For the present function the concept of 'orchestration' embraces the component activities: Differentiation (patterning), Time (i.e. absolute time) and timing, and fidelity (quality and risk management). These terms in combination are related to a strive for continuous improvement of the system in the right way and at the right time.

**Principle**: Survival of the fittest orchestration
**Mechanism**: Natural selection[15] of the fittest orchestration
**Emergent property**: Selected phenotypes

Note that we are still at a group level, building on top of the previous functions. According to Smaldino, "Group-level traits are possible when individuals display both *differentiation* and *organization*." ([73] *page 244*). By 'differentiation' he refers to individuals taking on different roles; and by 'organisation' he means differentiated individuals coordinating and collaborating for a shared purpose. This is the same meaning as the similar concepts in the template information model.

'Differentiation' is concerned with the establishment of patterns, therein speciation and specialisation.

'Time and timing' are implicit in all evolutionary activities – without time there will be no evolution – so in the context of the present function, the particular meaning refers to the explicit aspects of Time and timing in relation to evolution.

[15] Note the definition of 'natural selection' in Section 1.2.

'Fidelity' is concerned not only with how to ensure the quality of the phenotype at any given time, i.e. quality management, but also risk management. The existence of change is desirable – it is at the very core of an evolutionary theory, as it is the basis for promoting evolutionary advantages. So, risk management has to do with balancing between advantageous changes and detrimental changes, or preventing or compensating for detrimental changes, while also knowing that these may one day become advantageous.

| **Inevitable Evolution Theory (IET)** | **Standard Evolutionary Theory (SET)** | **Extended Evolutionary Synthesis (EES)** |
|---|---|---|
| Constituent properties in orchestration:<br>• Interspecies dynamics, including:<br>o Symbiosis and parasitism<br>o Interspecies cheating and deception<br>o Interspecific killing<br>• Time & timing<br>• Fidelity | Acquired properties in orchestration:<br>• Speciation<br>o Developmental symbiosis<br>• Specialisation<br>• Time & timing<br>• Fidelity | Adaptive properties in orchestration:<br>• Speciation<br>• Specialisation<br>o Eusociality<br>o Social patterning<br>• Time & timing<br>o Life-history theory<br>• Fidelity |

**Foreword to the Perspectives**

This chapter clearly shows that it can sometimes be difficult when analysing a given theory or observation to distinguish whether it belongs within the one perspective or the other, here in particular distinguishing the Standard Evolutionary Theory from the Extended Evolutionary Synthesis. The reason is that there may be insufficient details of how the underlying mechanisms exert the mentioned functionality/trait. For instance, the three factors referenced in Section 8.2.2 from Rueffler and co-workers' work ([120]), where bullet (c) clearly belongs in Standard Evolutionary Theory – or at least it is formulated like that, while bullets (a) and (b) could equally well belong in the Standard Evolutionary Theory and/or in the Extended Evolutionary Synthesis – the author suggests the 'and'.

It seems that the origin of eusociality evolved at least ten times in insects ([121]), and therefore it would be possible that some of them have involved the mechanisms characterising the Extended Evolutionary Synthesis, while others may have evolved primarily based on mechanisms from the Standard Evolutionary Theory – again a likely 'both/and' scenario.

## *8.1 Orchestration at Evolution, 'Inevitable Evolution Theory'*

### *8.1.1 Interspecies Dynamics*

In Section 6.1 the intraspecies dynamics was addressed, while the present section focusses on interspecies dynamics.

When this topic was judged relevant in this place in the Unifying Theory of Evolution, a small dedicated literature search on the topic was accomplished and it turned out that there were indeed a couple of highly relevant topics to address, as follows. For instance, it is suggested here that symbiosis between species as well as parasitism belongs here in the model.

#### *8.1.1.1 Symbiosis*

Sachs and co-workers review a series of examples of symbiosis and parasitism, the involved parties' mutual benefit from such relationship, and initiation as well as maintenance of such relationship ([122]), for instance:

- Ants and acacias: the acacia tree provides chambers and nourishes the ants that in turn defend the tree from animals getting in contact, with enhanced reproductive success for both parties: the ants' protection of the given plant against herbivores ensures the ants a future home and in return the plant provides a food supply in terms of protein and lipid rich bodies that nourish the ants living on the tree.
- Yuccas and Yucca moth: the moth pollinates the Yucca flower and in return moth larvae nourish on the fruit and seeds. There is a mechanism to reward moths providing low egg loads: "Through selective maturation of fruit with low moth egg loads and high pollen loads the plant has a partner choice mechanism to reward moths that do not overload plant ovaries with larvae (*refs*)." ([122] *page 150*).
- Squid light organ: the light is provided by a luminescent bacterium that the squids acquire selectively from the environment (based on a surface peptide on the bacterium). The squid's defence against non-luminous strains is to produce poisonous concentrations of peroxidase. In this case, the squid is taking advantage of the quorum response collective decision-making mechanism of the bacteria strain in question; see Ross-Gillespie & Kümmerli ([61]).
- Legume-rhizobial bacteria symbiosis, where the bacteria fix atmospheric nitrogen into organic forms that increase plant growth. A couple of mechanisms are involved in the symbiosis: i) the plant produces flavonoids that are specifically recognized and matched by some rhizobial strains to induce critical stages of the infection; and ii) cooperation is secured and maintained via a mechanism with chemical sanctions against non-symbiotic strains.
- Algal-invertebrate symbiosis, where the majority of algae are dinoflagellates and the hosts include a large variety of sponges, cnidarians, mollusks, flatworms and foraminiferans: the algae provide carbohydrates to the host and in return have access to other metabolic components needed plus a protected environment. The algae are acquired from the environment and maintained by the host via expel of excess algae and by a 'partner choice' mechanism (choosing cooperative partners) to acquire new algae. The mutual benefit scheme can operate at different levels of organisation and different timescales and is correspondingly vulnerable to exploitation.
- Leaf-cutter ants that cultivate gardens: The ant depends on the fungus for food and delivers food for the fungus in return. Associations between ants and fungal lineages can persist for generations – and hence for a prolonged evolutionary time – and clones may be transferred from a mother nest to offspring nests, although also lateral exchange may take place. The fungus, when associated with fungus-growing ants, is clonally propagated within the ant

nest. It seems that ants may be able to discriminate and select productive cultivars from unproductive ones through a 'partner fidelity feedback' mechanism, based on productivity measures; where the mechanism may merely refer to "*a coupling of fitness* between two individuals through repeated interactions (*ref*)." ([122] *page 140*).

Further, both insects and hummingbirds remember the characteristics of plants that provide rewards for visits, and hence selfish plant species have fewer visits ([122]).

*8.1.1.2 Parasitism*

Also parasites may affect the evolution and may do so by changing the state (behavioural, physiological, morphological, emotional or otherwise) of an infected animal. Considering that an infection may alter for instance the general well-being, the immune responses, susceptibility to secondary infections or reduced nutritional status, and even the reproduction (e.g., via an individual's reproductive behaviour, attractiveness as mates, and the host's sexual development), then it is obvious that parasite infections have a considerable potential to impact fitness of host animals; see Barber and Dingemanse ([123]).

Various fungi are specialised to invade plant tissue to exist inside of these plants in various places and ways (endophytic), and may cause plant diseases ([122]). The authors say that in general horizontally transmitted endophytes have deleterious effects on their hosts, consistent with a reduced or absent partner fidelity feedback between the horizontally transmitted endophytes and their hosts.

Barber and Dingemanse review in [123] the interplay between parasitism and the evolutionary ecology of animal personality, with known examples. They distinguish between personality ("consistent individual differences in the same behaviour across time and context ..." ([123] *page 4077)* and behavioural syndromes ("... distinct personality traits ... correlated within populations ..." ([123] *page 4077)*. Evidence suggests among others that:

1) Natural variation in behaviour differentially exposes individuals to different types of parasites, simply because the parasite infection may change one or more parameters of the host's behaviour, for instance, its patterns of habitat use, risk taking (e.g. boldness and exploration tendency), trappability, sociability, aggressiveness, and general level of activity;
2) Where parasites represent a significant threat, animals have developed a range of avoidance behaviours, as well as migration behaviour. For instance, regarding the former, neophilia (curiosity toward novel entities in the environment) predisposes the exploratory individuals to infection and therefore introduces a cost to exploration and similar for aggressive individuals. Therefore, parasites have the potential to impose a selection pressure similar to the behaviour toward predators;
3) "Parasite transmission strategies exploit a diverse range of host behaviours, including social, sexual and foraging behaviours." ([123] *page 4080*), for example, sexual promiscuity is a highly-significant risk factor for a wide range of sexually transmitted diseases;

4) Also, "... separate sympatric morphs typically develop divergent parasite faunas, reflecting morph-specific differences in foraging ecology and distribution (*refs*)." ([123] *page 4080*);
5) "To maximize their fitness value, behavioural changes that are brought about by parasites are predicted to be rather specific, influencing behavioural traits that facilitate transmission while leaving others intact." ([123] *page 4083*). For instance, it has been experimentally demonstrated that infected freshwater gammarids display altered responses to predator stimuli, caused by a modification of the host's serotonergic pathways – that is, alteration of a neurotransmitter level in the host.

It seems highly likely that parasites affect the evolution of their hosts through a direct effect on the host's fitness and likely even to the extent of an arms race between viruses and their hosts that according to Koonin and Wolf may be one of the defining factors of evolution; see [24].

#### *8.1.1.3 Interspecies Cheating and Deception*

The exhaustive and entertaining paper of Ghoul and co-workers on cheating in general illustrates the various kinds of cheating, including deceit, within and between species – see [98] and Section 6.2.1.1. For this perspective on interspecies dynamics, it is specifically the cheating and deception that are in focus. Deceit/deception without cheat is explained as "deception will not be cheating when it is exploiting a feature of the world that can be used as a guide to action, termed a "cue" (*ref*)." ([98] *page 325*). An illustrative example is the anglerfish that lure in prey with bioluminescent bulbs; anglerfish are deceptive but not exploiting a cooperative signalling system. An example that the authors bring on deceptive cheating is the fork-tailed drongo that manipulates pied babblers or meerkats by making false predator alarm calls, thereby enabling the drongo to steel their food; this manipulative strategy works in small groups of the pied babblers where there are fewer guards, but not in larger groups where more individuals participate in predator vigilance ([98]).

There are several examples of cheating between species, some of which involve deceit, like the cleaner fish that takes a bite of the client rather than cleaning off the parasites once it gets close enough ([98]). Other of Ghoul and co-workers' examples in this respect are for instance the Yucca moth that lays eggs in the fertilised and developing Yucca fruit without pollinating the flower, and plants that have flowers without nectar and still attract pollinators with scent, morphology or colours.

#### *8.1.1.4 Interspecific Killing*

Donadio and Buskirk define 'predatory habit' as "the degree to which a carnivore kills and consumes vertebrate prey." ([124] *page 524*). The interspecific killing is likely to affect the exploitative competition and also influence habitat use and distribution of competing species. The authors conclude from their study, where they pairwise compare a large set of carnivores, that frequency, intensity and direction of killing interactions depend on factors like overlap of food sources, relative body size, trophic relationships, and taxonomic relatedness. The overlap of food points at competition for food as the motivating factor. The relative body size points at the minimization of risk as a modifying factor. Predatory habits influenced the occurrence of lethal encounters, and "for closely related species it is more difficult to avoid risky confrontations." ([124] *page 533)*.

Hoogland and Brown in [125] report that the white-tailed herbivorous prairie dog reduces interspecific competition within its habitat by killing its competitor, the herbivorous Wyoming ground squirrels without eating it, which is unusual at interspecific killing; this therefore clearly points at the competition for food as a motivational factor.

##### *8.1.1.5 Miscellaneous*

Other candidate topics that might be considered as belonging in this position in the Unifying Theory of Evolution but which are not discussed in detail are:

- Differentiation with respect to food resources may ease interspecies competition for resources and thereby also ease their co-existence when there is not a predator-prey relationship
- Interspecies cooperation/altruism

Domestication of other species for one's own purpose (like human's agriculture, ants milking aphids, and similar examples) is referred to Chapter 10 because there is more than a simple cooperation involved, as found for symbiosis and parasitism.

### *8.1.2 Time and Timing*

As seen from Table 1, this topic was judged relevant here in the model, simply because Time and timing are an essential element of the concept of 'evolution'.

Witting in [22] discusses the concept of 'evolutionary arrow of time'. By 'evolutionary arrow of time' he (and similarly Ekstig in [126]) refers to directionality of evolution toward constantly increasing complexity. Witting mentions the analogy of the 'evolutionary arrow of time' with the entropy of an isolated thermodynamic system where the entropy can only increase with time ($2^{nd}$ Law of Thermodynamics).

Kapeller and co-workers, using Witting's theory, cite literature findings for another aspect of time – that is, the development time of an organism (i.e. growth rate, generational time); see [65]. For instance, the nutritional environment of the parental generation influences the development time, fecundity and dispersal potential of offspring, and moreover, pupal size positively affects the larval survival of the offspring generation. The authors among others conclude that 1) for the model, it doesn't matter whether the density-dependent changes of growth rate are caused by plasticity or different genotypes (balanced polymorphism), and 2) that the population dynamics of *T.viridana* according to Witting's model are influenced by two forces that together may be viewed as a development-reproduction trade-off mechanism: a) enhanced population growth rate when the population density is low, and b) the benefit of competitive traits when the population density is high.

Further, Benton and co-workers emphasize that "Population structure directly influences population dynamics because of the time it takes for the organisms to complete the life cycle." ([64] *page 1176*), because different stages may respond differently to changes in the environment, and also because time lags in general may affect the dynamics. That is, generation time and cycles may create asymmetry in cohorts, which has been shown for many well-studied model organisms ([64]).

Another candidate timing topic illustrating the relevance of Time and timing for evolution is synchronisation of spawning in corals; see Sorek and co-workers in [127]. Reproduction in corals is influenced by three main types of cycles: 1) time (annual / seasonal rhythm related to sea temperature), 2) monthly lunar rhythm, and 3) "… the diel light cycle, which acts as a zeitgeber ("time giver") and induces spawning to occur after a precise period of darkness…" ([127] *page 51*). The evolutionary advantage of such timed mass-spawning events is the reduction of overall predation, but of course there is then the risk of possible formation of non-viable, interspecies hybrids ([127]).

### *8.1.3 Fidelity*

Fidelity was judged relevant as a factor in evolution, but little dedicated literature was found on the topic specifically in relation to evolution. Nevertheless, some aspects of relevance were identified in relation to fidelity for the Inevitable Evolution Theory, namely some problems identified for 'Quorum sensing' that is previously discussed in Section 7.1.

Quorum sensing has a built-in problem of inaccuracy in the discrimination, reported among others by ([101] *page 1266*), namely that "With respect to quorum sensing, increasingly inaccurate quantity discrimination with increasing stimulus magnitude means that individuals are more likely to misjudge the quorum point as quorum size increases.". There is a risk that an incorrect decision will be reached if initial decisions are poor and influence subsequent decisions in a sequence of decisions, or even a single wrong decision. The authors say that "In general, lower quorum thresholds are associated with more rapid, and less accurate, decision-making …, whereas higher quorum thresholds occur in more benign environments and larger colonies …" ([101] *page 1266*). Such problem of inaccuracy may arise simply due to small variations within the population, while the higher threshold for larger populations will compensate for such variation at low thresholds.

Franks and co-workers in [128] repeat the mentioned experiments of Cronin, but now with fluctuations in quality for one of the candidate new nest sites, showing that the ants by means of the quorum sensing have the ability to estimate the average value quite accurately of a resource that fluctuates in quality, even over a relatively prolonged period. This shows that they are able to handle conflicting data such as a single poor rating amongst a number of excellent ones, and they can even employ change in response to environmental conditions and the need for urgency. This confirms the statement of Cronin suggesting that "ants may employ higher quorum in more complex choices…" in connection with finding new nests ([101] *page 1267*).

We see these abilities as approaches to optimise quality of the decision making and to handle risks, and once a paradigm mechanism is established for either, evolution will find ways to further refine and/or exploit such mechanism.

## *8.2 Orchestration at Evolution, 'Standard Evolutionary Theory'*

### *8.2.1 Speciation*

Interspecific differentiation, speciation, is the evolutionary process establishing new species. Or, as Seehausen and co-workers formally define 'speciation': "we define speciation as the origin of reproductive barriers among populations that permit the maintenance of genetic and phenotypic distinctiveness of these populations in geographical proximity." ([129] *page 176*).

The review of Seehausen and co-workers, [129], that was initiated at a European Science Foundation workshop is extremely dense with information on speciation and also with suggestions for mechanisms supporting speciation. The authors distinguish between 'intrinsic postzygotic isolation' ("Fitness reduction in hybrids that is independent of the environment." ([129] *page 178*)) and 'extrinsic postzygotic isolation' ("Fitness reduction in hybrids that is dependent on the environment and that is mediated by genotype-environment interactions." ([129] *page 178*)), prezygotic isolation ("Effect of barriers that act before or after mating but before fertilization, including the effects of divergent mate choice, habitat preference, reproductive timing and gametic incompatibility." ([129] *page 180*)), and sexual isolation. From these definitions, it is obvious that the intrinsic postzygotic isolation belongs in this perspective, while the extrinsic postzygotic isolation belongs in the perspective of Extended Evolutionary Synthesis, and therefore is referred to Section 8.3.1.

According to Seehausen and co-workers, divergent adaptation rarely causes sufficient reproductive isolation on its own to enable the accumulation or the tenacity of species differences in geographical proximity ([129]). It seems to be the interplay between the intrinsic and the extrinsic postzygotic isolation that established the creation of new species, and therefore, the further description of the interplay is referred to Section 8.3.1.

Seehausen and co-workers say evidence suggests that the most frequent cause of intrinsic postzygotic isolation is negative epistatic interactions[16], but that also other mechanisms may lead to this phenomenon, such as underdominance[17], gene duplication, transposition and gene loss. Further, genomic conflict[18] may be a common mechanism behind intrinsic postzygotic isolation, arising as a result of competing interests of males and females, from meiotic drivers, mobile elements or other 'selfish' genetic elements and their suppressors, "and from competing interests between genomes of organelles and the nucleus." ([129] *page 180*). The authors further explain that "For intrinsic isolation, incompatibility factors that are driven by genomic conflict are expected to accumulate in genomic regions of reduced recombination …" under certain conditions ([129] *page 180*), and that "Sex chromosomes are particularly susceptible to the accumulation of incompatibility factors that are derived from genomic conflict because these chromosomes are constantly in a 'battle' over segregation, whereas only small and tightly linked autosomal regions are in conflict with their homologues[ref]." ([129] *page 180*). The authors conclude that genomic conflict may be a frequent source of intrinsic postzygotic isolation. Finally, the authors conclude that for sympatrically co-existing species (populations/species sharing habitat, so they may exchange genes) the differentiation is scattered across the genome, while the adaptive differentiation between parapatric populations (populations/species living separated, yet adjacent habitats) may be restricted to a few genomic islands.

Frías in [130] suggests a supplementary mechanism whereby speciation may happen, without geographic barrier (reproductive isolation) or parapatric speciation:

---

[16] Epistasis is "the interaction between two or more genes at different loci such that phenotypic expression of one depends on expression of another." ([10]).

[17] Underdominance refers to "Heterozygotic inferiority, that is, the phenotype expressed in heterozygotes has lower fitness than that of either homozygote. This can cause disruptive selection." ([129] *page 181*).

[18] Genomic conflict is defined as the "Conflict that arises between genes or genetic elements within the same genome either when they are transmitted by the same rules (for example biparental versus uniparental inheritance) or when a gene causes its own transmission to the detriment of the rest of the genome. …" ([129] *page 178*).

occurring via intrinsic barriers caused by chromosomal rearrangements and negative heterosis (change in function in hybrids offspring). This could align with the finding of Mendez and co-workers in [131] that the Neanderthal Y-chromosome cannot be found in the human genome even if we all carry a small fraction of the rest of the Neanderthal genome. The authors suggest that "It is possible that incompatibilities at one or more of these genes played a role in the reproductive isolation of the two groups." ([131] *page 728*), since "Polypeptides from several Y-chromosome genes act as male-specific minor histocompatibility (H-Y) antigens that can elicit a maternal immune response during gestation." ([131] *page 732*).

A couple of recent papers presented in the below text show examples where major evolutionary steps seem to arise from changes in a relatively small number of genes. Such theories may all be valid in an evolutionary context, and should not be mutually exclusive; and apparently without the interaction with the environment.

First, recently Kronforst and co-workers in [132] revealed that 1) the origin of new butterfly species from old ones involves only a small number of genes, followed by interspecific gene flow, and that 2) the divergent genomic regions display evidence of both selection and adaptive introgression[19].

Another recent paper by Iskandar and co-workers, [133], presents a fanged frog recently discovered in Indonesia. Even if it is not an example directly concerned with speciation directly it does point at a small number of changes being able to account for major evolutionary transitions. Iskandar and co-workers present a frog species, which give birth to tadpoles without any intromittent organ present to facilitate sperm transfer: Out of 6455 known frog species, fewer than a dozen are known to have internalised fertilization, and all of these but the new species either lay fertilized eggs or give birth to froglets. The presence of a large yolk reserve in the birthed tadpoles may indicate that they are not yet fully matured for an independent life. Further, it seems that it would be a larger step to succeed with internalised fertilisation when there is no organ(s) to facilitate sperm transfer. More extreme withhold of the tadpoles would lead to giving birth to froglets, while a yolk reserve makes a delayed birth feasible. Each of these steps seems plausible, and the birthing of tadpoles therefore might be a missing link showing the evolution of the progeny state at delivery from an egg to a fully developed baby frog. Might it be reasonable to suggest that giving birth to tadpoles would require a small number of otherwise inconsequential steps related to delaying ejection of the fertilised eggs?

Hence, it cannot be excluded that Darwin in some cases was correct that small steps in the long run may suffice to feed mechanisms of evolution even at the level of generating novelty, without assuming that this takes place in a continuous smooth process. This may have been one interim mechanism for speciation until epigenetic mechanisms for interaction with the environment entered the scene of biological evolution.

#### *8.2.1.1 Developmental Symbiosis ('Eco-Evo-Devo')*

A relatively new field within evolution theories is called 'Eco-Evo-Devo' (Ecological Evolutionary Developmental Biology), seeking to embrace topics like horizontal transfer of genetic material (also called 'developmental symbiosis'), developmental plasticity, genetic accommodation, extragenic inheritance and niche construction; see Gilbert

---

[19] Introgression refers to the gene flow from one species into another by repeated back-crossing of an interspecific hybrid with one of its parent species (definition inspired by Wikipedia, 8th of June 2016).

and co-workers in [134]. Their review focuses primarily on developmental symbiosis and developmental plasticity, for which at least for the former belongs at this place in the Unifying Theory of Evolution, because of the suggested contribution to speciation.

Gilbert and co-workers define developmental symbiosis as follows: "Developmental symbiosis is the concept that organisms are constructed, in part, by the interactions that occur between the host and its persistent symbiotic microorganisms." ([134] *page 611*). The authors say that developmentally active symbionts provide mechanisms for the reproductive isolation mentioned in Section 8.2.1 above and in 8.3.1, and that the multispecies organism (called a holobiont or a metaorganism) is a dynamic community in or on a host organism integral to the functionality of that host's cells and organs (and vice versa). Such a strong statement demands an explanation based on documented evidence: a) "bacterial symbionts are essential for the metamorphosis of many invertebrates[refs], for the formation of ovaries by the wasp *Asobara*[ref] and for the germination of orchids[ref]. Moreover, the anterior-posterior axis of the nematode *Brugia malayi* is generated with the help of *Wolbachia* bacteria, …[ref]." ([134] *page 612-3*); b) gene expression profiles of the germ-free mice and zebrafish compared to the normal controls suggests that the microbiota significantly changes the expression of genes involved in cell proliferation, nutrient utilization and immune function (all connected with the completion of gut differentiation) as well as haematopoiesis, and in general, germ-free animals have serious immune system defects ([134]).

More convincing as evidence are the experiments studying the basis for the contribution to speciation, which have been accomplished for three related wasp species and have identified the gut microbiota causing the reproductive isolation: "The wasp species *Nasonia giraulti* and *Nasonia longicornism* have a similar range of gut bacteria and can produce healthy hybrid offspring, but when either wasp mates with the more distantly related *Nasonia vitripennis*, which has a different gut microbiota, their hybrid offspring die." ([134] *pages 615-6*). However, when such hybrid offspring of *N. vitripennis* with one of the other two wasps are raised in a germ-free environment, their offspring survive, and when the germ-free offspring of *N. vitripennis* are inoculated with the microbiota of the other two wasps, their offspring die.

A further example of the symbiont-induced reproductive isolation is concerned with the mating preference of *D. melanogaster*, which shows a strong mating preference for individuals raised on the same diet; however, such mating preference disappeared after antibiotic treatment, and reappeared after inoculation with microorganisms from the dietary media. The changes in mating preference were linked to the presence of *Lactobacillus plantarum*, "which was found to alter the cuticular hydrocarbons that form part of the mating pheromones of the adult fly." ([134] *page 616*).

In retrospect, because of a homeostatic relationship between the symbiont and the host that has mechanisms for co-development (mechanisms for recognising and maintaining complex communities of beneficial microorganisms) and co-evolution, one could see the developmental symbiosis as an advanced kind of multi-level selection.

### *8.2.2 Specialisation*

Specialisation (intraspecific differentiation) – that is, dividing tasks/roles on individuals. Specialisation is one way of maximising efficiency and effectiveness through division of tasks on roles.

Ross-Gillespie and Kümmerli express a point of relevance for the discussion on specialisation: "In effect, segregation into informed "leaders" and uninformed "followers" could facilitate a division of labor, with potential gains in efficiency at the group level." ([61] *page 8*).

Aplin and co-workers in [112] demonstrate that this is the case for great tits with variations in proactive versus reactive behaviours: "… if animals differ in their degree of sociality, then variation in the strength of social cohesion may mediate group-level movement, with asocial animals exerting directional 'pulling power' on more social individuals [*ref*]. Such emergent leader (initiator)–follower polymorphisms [*ref*] have been observed in the grouping behaviour of a diverse range of taxa, and leadership tendencies are increasingly proving to be consistent and repeatable within individuals [*refs*]." ([112] *page 1*). Further, the authors say "Exploration behaviour is repeatable [*ref*], heritable [*ref*], under selection [*refs*] and linked to a range of life-history traits in several populations [*refs*]." ([112] *page 2*). Studies show that mixed personality colonies are more successful in ants and social spiders, and thus, this variation might be the basis for the evolution of specialisation ([112]).

Smaldino emphasises the concept of 'organisation', which he perceives as differentiated individuals coordinating and collaborating for a shared purpose, while emphasising that such organisation itself is not the group trait ([73]). He further emphasises the following:

1) "Group-level traits are possible when individuals display both *differentiation* and *organization*." ([73] *page 244*) – by differentiation, he means that individuals take on different roles. And importantly in the present context of biological evolution, "A group-level trait is the *phenotypic effect* of social organization." ([73] *page 244*);
2) "Human groups organize in ways that produce emergent group-level traits. … Group-level traits are not expressed by any single individual in the group, but emerge from the structured organization of differentiated individuals." ([73] *page 244*). Moreover, "The difference between aggregate and emergent properties is often relative, but a useful heuristic for distinguishing group-level traits from collective behaviors is that the latter depend strongly on the specific organization of differentiated individuals, whereas the former do not." ([73] *page 245*);
3) Patterns of interactions and cooperation are not at random. "Interdependence sustains cooperation and provides a stable environment of mutual aid in which differentiation, division of labor, and complex group organization can emerge." ([73] *page 248*).

Lehmann and co-workers in [135] in a simulation experiment show that the evolution of sterile workers can easily happen under minimal life-cycle conditions known to promote altruism, such as: "… the possibility of kin recognition, punishment of defectors within groups, long-lasting niche construction effects, different modes of dispersal and competition between groups, different life histories such as overlapping generations and non-Poisson progeny distribution ..." ([135] *page 1891*).

Rueffler and co-workers in [120] present a unifying framework for the evolution of functional specialisation of repeated modules covering the range from multicellular organisms to advanced organisms. They illustrate this with a series of examples of functional specialisation in organisms ranging from multicellular bacteria and algae, to plants and on to organisms with higher levels of organisation, accounting for asymmetry in bilateral organisms, colonial organisms, and onto eusocial insects in colonies. Their conclusion is that division of labour is not an inevitable outcome of evolution, and therefore they derive a framework with a set of minimal requirements for division of labour to be favoured by selection. They show that evolution of the division of labour is favoured by three factors with the caveat that these are necessary but not necessarily sufficient alone for division of labour to evolve by natural selection:

1) Positional effects, which may include asymmetries. For instance, the limbs at an anterior segment of an arthropod are more likely to contribute to the task of food processing than a posterior limb, and similarly anterior teeth are more likely to contribute to cutting food than posterior ones;
2) Accelerating performance functions for the case of two equivalent modules: where "the gain in performance through increased specialization of one module exceeds the loss due to decreased specialisation of the other module, resulting in increased performance at the level of the whole organism." ([120] *page E329*);
3) Synergistic interactions between modules or gene products in general favour specialisation. Interactions between gene products coded by different loci can emerge for instance when enzymes are dimers or multimers. In the case of duplicated genes where the two loci accumulate different mutations (achieving heterodimers) and such dimers perform better than the original homodimers then synergy is achieved and specialisation is favoured at selection. Rueffler and co-workers suggest that duplicated genes are easily maintained if degenerative mutations affect complementary regulatory regions; under such conditions duplicated genes are not only maintained but also become specialised for alternative tasks. See also the theory 'Constructive neutral evolution' in Section 9.3.2.

That eusociality has evolved at least ten times in insects ([121]) makes it plausible that specialisation in an evolutionary context is a favourable trait; might this be because specialisation puts lesser demands on the range of skills necessary for each individual at increasing complexity, as suggested by Rueffler and co-workers saying "As a consequence, performance increases when half the modules specialize for one task while the remaining modules specialize for the other task." ([120] *page E332*).

Task sharing has been found to increase reproductive output for a cooperative breeding bird; if such task-sharing benefits make a group more efficient, evolution should lead to behavioural consistency ([74]).

Different social network structures may provide opportunities for different social roles within and between groups; for instance bottlenose dolphins have brokers facilitating interactions between groups ([74]), yet the mechanism for how individuals obtain such role and how this affects the cooperative climate within the group is not known.

In summary, it seems that the Standard Evolutionary Theory can make a full account for the evolution of eusociality, and there cannot be any doubt that genetics is a major factor in this evolutionary process.

### 8.2.3 Time and Timing

According to the template information model (see Section 14.1), the 'Time and timing' is a central topic in the Function 5, and therefore, it is of relevance to see which particular Time and timing issues are involved in this perspective.

Suggestions for 'Time & timing' aspects of relevance here are addressed in simulation studies, explicitly or implicitly, such as:

- Mutational frequency. 'Frequency' is the number of appearances per time unit, and since mutations – of all kinds – are a basis for evolution it is obvious that mutational frequency is a factor in evolution. The number of mutations is far from scarce, and repair mechanisms exist for most, so that mutations will not run amok.
- Generation time is the pre-reproductive period – that is, the length in time from 'birth' (or delivery of a fertilised egg) to onset of the reproductive period. This is extensively discussed for humans in the section on Life-History Theory, but also works for other species; see Section 8.3.3.1. It is obvious that the duration of the biological development from an egg into a mature reproducing being is of importance as an evolutionary factor, since every day and every minute bears the risk of being de-selected, for instance by illness, by predators, by starvation, and more.
- Length of the reproductive period, and thereby also the number of cycles of reproductive activities and processes; the variation is tremendous: some reproduce continuously (e.g. some queen bees), others only once and then die, and others in repeated cycles, some only once per year or only in a specific season of the year.
- Whether there are overlapping generations. Overlapping generations open for the opportunity of parental care and in some cases even for older siblings to help in the upbringing, as is the case with some birds and also humans. The lack of overlapping generations puts special demands on the robustness of the biological development, as the progeny has to be completely independent and self-managed from birth/hatching.

Further, Kirschner in [7] argues that biological systems are able to accumulate (non-lethal) variation, which then will penetrate in the phenotype at some point in time. Kirschner here refers to calendar time. The present author further suggests that biological systems are able to accommodate (non-lethal) variation, when this penetrates in the phenotype at a point in life where it is insignificant as a factor in the evolutionary selection, for instance at senescence (i.e. timing of events). One such example is the age-related macular degeneration (AMD – causing blindness), which is clearly non-lethal and shows up after the reproductive period. That AMD shows up late in life implies that it will not be a fitness factor in the genetic evolutionary pressure influencing the capability of generating viable and competitive progeny or influencing the propagation of such. Therefore, it has been feasible for it to 'hide' and express itself only when people get old or when it is exposed as a result of other biological conditions. This kind of time-dependent visibility of genomic changes must be valid for advantageous as well as detrimental aspects of a phenotype, and is known from the concept 'onset' of clinical symptoms.

Another similar example for humans is Dupuytren's Contracture, which is an inherited proliferative connective tissue disorder that primarily affects the hands; it worsens gradually after the age of 40 and decreases the patients' ability to hold objects. Since the syndrome usually only develops enough to be diagnosed after the age of 40, most people have ended their reproductive career, at least in the western cultures, and therefore, the effect on reproductive fitness is insignificant and only indirect via the support of care by grandparents to their children and grandchildren.

### *8.2.4 Fidelity*

A conclusion in [1] was that nature has indeed provided mechanisms to handle every little trouble-maker within the system – that is, to handle risk and manage quality of the system.

An example of risk management is the reduction of the effect of a disadvantageous evolutionary change that came through the evolution of polyploidy[20]. Having more than one allele[21], while for instance forcing the zygote developmental process to choose among the candidate alleles, provides nature with an option for survival through hiding or promoting a specific gene at random or coordinated with other genes on the same allele for a given new phenotype. This means that a disadvantageous genotype may survive (e.g. as a recessive change versus a dominant change) until conditions for it are opportunistic.

Examples of other fidelity mechanisms under the Standard Evolutionary Theory are fidelity in relation to the mutation processes behind evolution, for instance:

- Repair mechanisms, e.g. for damaged DNA that arose under the processes establishing evolutionary changes. A conclusion in [1] was that whatever can go wrong has a repair mechanism.
- Risk management is also concerned with preventive initiatives. From this perspective, one such mechanism is multiplication of genes (establishing redundancy), which 1) gives the opportunity of having a reserve while 'experimenting' with changing the other to achieve new/refined functionality (see also the discussion in Section 9.3.1); or 2) enabling a larger gene expression range. Redundancy appears in various ways: a) redundancy of genes, where a multiplum of genes provide resources for performing the same task, but with small and tolerated variation that may serve to buffer or oppositely promote the new functionality of that phenotype; b) redundancy in the genetic

---

[20] From [1]: 'Ploidy' refers to the number of sets of chromosomes in a given cell type. 'Euploidy' refers to the cell having an equal number of all chromosomes, excluding the sex chromosomes. 'Aneuploidy' refers to a cell not having euploidy, as for instance in Down's syndrome where there is an extra (trisomy) of chromosome 21. The majority of human somatic cells are diploid (having two sets of chromosomes = 2x23), i.e. they have one copy of the mother's and of the father's genetic material. The term 'haploid' is used for the chromosome set in a gamete (ovum and sperm). In humans, the haploid number is 23 – that is, the human gametes are 'monoploid'. When the ovum is fertilised by a sperm, the zygote is formed as the first diploid cell of a human being.
'Polyploidy' (multiploidy) in the context of humans is used to describe cells having more than the normal diploid set of chromosomes. There are normal polyploid cells in the human body; these are formed through either several rounds of DNA replication without cell division (endomitosis, e.g. hepatocytes and megakaryocytes that produce blood platelets) or fusion (e.g. myocytes and cardiomyocytes). In contrast to polyploidy, erythrocytes have lost their nucleus and hence are 'nulliploid' cells.

[21] Allele, any of the forms of the same gene that occurs at the same locus on a homologous chromosome – i.e. one of the versions in diploid or polyploid genes.

code in that multiple codons may code for the same amino acid – because of the wobbling base pairing; for instance, there is a lot of redundancy for tRNA, but the tolerance for changes is small.

- Genomic drift has the potential to increase efficiency through coordination of genes that with benefit are expressed at the same time.
- History of evolutionary change: since mutational changes accumulate over time and are in principle stochastic in nature (i.e. independent of the phenotypic consequence), alternative phenotypic outcomes depends on the order of appearance of the mutations ([26]).
- 'Bet hedging': comprises a principle of stochasticity applied to exploit mutational heterogeneity and thereby reducing the risk of evolutionary dead ends. This is suggested in the hypothesis paper, [136], simply taking advantage of the initiation of alternative solutions each with different mutation-based risks associated. "In a heterogeneous environment, bet-hedging is favoured when an increase in the phenotypic variance means that at least some individuals will express the phenotype with the maximal fitness." ([27] *page 506*).

## *8.3 Orchestration at Evolution, 'Extended Evolutionary Synthesis'*

The basic mechanism behind Extended Evolutionary Synthesis is adaptation to changing environmental conditions, regulated through plasticity.

### *8.3.1 Speciation*

Speciation is the process at which a population of biological individuals evolves to become distinct new species.

The conclusion in the study of Simola and co-workers is that "evolutionary changes in gene regulation seem to dominate our view of the shared genomic features associated with the origins of eusociality." ([137] *pages 1243-4*). Among others, their results show that "changes in the transcription factor (TF) repertoire were important in the initial stages of ant evolution." ([137] *page 1238*). The regulation of gene expression as a key factor in the evolution of eusociality is pervading in their study results; for instance, "These results suggest that caste-associated gene expression plasticity is a continuously evolving trait in eusocial insects that is partly determined by TFBS abundance …", and "We were struck by the over-representation of TFs associated with eusocial regulatory evolution (…) among key regulators of these networks…" ([137] *page 1241*).

Since phenotypic plasticity is capable of hiding genetic variation, it is capable of promoting the accumulation and release of cryptic genetic variation – that is, variation that is only expressed under certain conditions ([47]). Thereby, because cryptic variations are unexpressed, potentially a large set of genetic variants may be hidden from selection, while still allowed to drift and accumulate in natural populations, independently for the individual genetic components affected ([55]). The mechanisms stem from the buffering of novel genetic variants by compensatory plastic responses, but also because environment-specific genes are subject to relaxed selection in the non-inducing environment. This is along the line of reasoning argued in [47] and [55] for how plasticity as a mechanism can facilitate / support / enhance the process of speciation. Pfennig and co-workers nicely summarize the evidence highlighting plasticity's role in diversification and speciation in their Box 3 as follows. First the direct evidence:

- "plasticity can mediate rapid and adaptive divergence between populations …(*refs*)
- plasticity in traits that influence mate choice (*ref*), resource or habitat use (*refs*)…, or phenology (*ref*) can promote rapid reproductive isolation
- clades in which resource polyphenism[22] has evolved are more species rich than sister clades (*ref*)
- the occurrence of homoplasy in conditionally expressed traits …(*refs*)
- the prevalence of replicated adaptive radiation involving environmentally induced traits …( *refs*)"

(end of citation from ([47] *page 464*)

They continue by listing phenomena that can be illuminated by considering plasticity:

- "maintenance of cryptic genetic variation (*ref*)
- peak shifts on adaptive landscapes (*ref*)
- origins of novel traits (*ref*) and body plans (*ref*)"

(end of citation from ([47] *page 464*)

The review of Seehausen and co-workers give a detailed account of the mechanisms behind speciation and a lot of supplementary details: In speciation driven by divergent selection (i.e. "Selection that favours different phenotypes in different populations." ([129] *page 178*)), "extrinsic postzygotic and prezygotic barriers evolve first and often interact to produce reproductive isolation, and intrinsic postzygotic barriers will often only evolve later in the speciation process …" ([129] *page 179*) (see the definitions of these concepts in Section 8.2.1), leading toward the end result of two irreversibly isolated species. The other alternative, "speciation driven by intrinsic barriers often results from epistatic incompatibilities, which may … accumulate in an accelerating 'snowball' manner[refs] either as a by-product of selection or as a result of genetic drift (which only occurs slowly). Extrinsic postzygotic and prezygotic barriers may accumulate later…", and this combination again may lead toward the end result of two irreversibly isolated species ([129] *page 179*). So, it is the combination of the adaptive element and the genetic foundation that leads to the generation of new species, in a speciation continuum. Seehausen and co-workers say that both prezygotic and extrinsic postzygotic isolation often evolve faster than intrinsic postzygotic isolation.

Further, Seehausen and co-workers express that adaptive divergence accumulate preferentially in regions of low recombination, including "the centres of chromosomes[ref], the vicinity of centromeres[ref], sites of inversions[ref] and often … sex chromosomes[refs]." ([129] *page 183*).

A supplementary thought: From the perspective of physics, Bak and Boettcher in [28] specifically discuss the issue of small evolutionary steps versus major evolutionary

---

[22] Polyphenism: a mechanism enabling multiple (discrete) phenotypes to develop from a single genotype as a result of different environmental conditions.

impact. They express that "Systems in equilibrium are linear, so the underlying picture is one where nature is in balance. However, in physics we are aware that many dynamical systems show non-equilibrium, non-linear behaviour. In particular, large dynamical systems are known to evolve to critical states, where the response to small impacts may be enormous, reflecting a divergent susceptibility." ([28] *page 143-4*). According to their theory, the split of a single species into two distinct species is not a steady and gradual (continuous) transformation. The visible behaviour of systems displays a temporal and/or spatial scale invariance characteristic at the critical point, which leads to a transition. Generating novelty in a single step seems implausible, because this would require a (large) number of coordinated changes, while changes that either are latent or inconsequential for the phenotype may accumulate until the external conditions favour such characteristics or abilities. Thereby, punctuated equilibrium may contribute to major evolutionary steps. If so, then we are all carrying the potential for novelty, and various cases of genetically-based disorders may from an evolutionary perspective be perceived as mishaps of such changes or perhaps merely a bad timing for their penetration in the phenotype.

Might it be that the 'snowball' manner mentioned by Seehausen and co-workers could be carried into effect by the punctuated equilibrium? One of the differences between the two mechanisms explained by [129] and the suggestion in [28] – that punctuated equilibrium may contribute to major evolutionary steps, which includes speciation – is whether or not the speciation processes represents a continuum. If the answer to the question is 'yes' then both parties might be right, namely that the initial accumulation of relevant changes may be a continuum, and that a certain point the punctuated equilibrium takes over and completes the speciation process.

### *8.3.2 Specialisation, Eusociality*

Eusociality is "an advanced state of sociality found primarily in insects and characterised by reduced reproduction in workers that care for offspring other than their own (*ref*)." ([138] *page 1*). The adaptive element in the various eusocial phenotypes being determined by the food clearly indicates that 'eusociality' belongs at this place within the Unifying Theory of Evolution.

Eusocial species are animals living in colonies formed by family groups ([139]). Eusocial species are found wide-spread in nature, most common among insects such as species of ants, bees and termites, but also include thrips and aphids as well as a shrimp species and a naked mole rat ([139]).

The colonies of eusocial species are collections of highly genetically related individuals, and "… the emergence of group-level traits in these species is encoded at the level of the genotypes of the foundress queens, their mates, and their offspring." ([73] *page 248*). They are characterised by a reproductive division of labour and a clear distinction between members and non-members, of which the latter are excluded from entry into the nest or territory ([139]). This, which reminds one of the concept of greenbeards (see this concept in Section 5.2.3), makes a mechanism for recognition of membership an essential factor in evolutionary success in the present context. Evidence from eusocial insects suggests that this discriminative mechanism is based on chemical cues in various ways, like a dedicated gland, or the queen secreting a unique labelling mixture that is shared by rubbing, or shared nesting material ([139]); in any case the necessary variation appears analogous to the immune system of vertebrates. It further requires a mechanism for acting in ways that exclude non-members from the colony

([139]), which brings an association to the concept of punishment of cheaters in altruistic and cooperative behavioural patterns.

Over time, even subtle differences in abilities and circumstances can give rise to differentiated individuals who are dedicated in their roles and deeply specialised ([73]). Throughout the development of an individual, these are drawn to different roles based on opportunity, experience and epigenetic predispositions – that is, 'task-aptitude' as discussed by Ibbotson in [140]. Specialisation implies that the individual may itself evolve within its role.

Bergmüller and co-workers suggest that negative frequency-dependent processes resulting from social competition in a multi-niche social environment may lead to individual variation in social roles and associated personality types, thereby generating individual differences in behavioural strategies ([74]). Moreover, "…helping behaviour should be regarded as a behaviour that is a mixture of constraints imposed by ontogeny and phylogenetic heritage and an adaptive response to particular selective pressures (*ref*)." ([74] *page 2756*). "Adaptive behavioural correlations can result if multiple ecological or social challenges favour particular trait combinations." ([74] *page 2756*), and it seems to be the combinations of traits that are subject to selection. See also Section 8.3.2.1 in this respect.

Two papers bring an extraordinary insight into the evolution of eusociality: [138] and [141]. Nowak and co-workers excellently summarise their model of the evolutionary progression as comprising the following series of stages ([141] *page 1062*):

a) The formation of groups;
b) The groups are tightly formed through the manifestation of a minimum and necessary combination of pre-adaptive traits;
c) The appearance of mutations that prescribe the persistence of the group, most likely by silencing of dispersal behaviour. Therefore, a durable nest remains a key element in maintaining the prevalence. From this stage, primitive eusociality may emerge due to spring-loaded pre-adaptations;
d) Emergent traits caused by the interaction between group members are shaped by environmental forces combined with natural selection;
e) Multilevel selection continues to drive changes in the colony's life cycle and social structures, which may lead to extremes.

Hunt in [138] emphasises a similar, but a little elaborated and convincing model, also with incremental steps and each step characterised by its own evolutionary mode, as was Nowak and co-workers' model. The model contains "four grades of social organization, three modes of evolutionary change, two thresholds, and a gradient of selection." ([138] *page 5*). There are two maternal types: 'allomaternal' (caregivers of the offspring without being the biological mother) and 'maternal'; and the sequence of the four stages includes: solitary (non-eusociality), facultative eusociality, primitive eusociality, and advanced eusociality. The shift from the solitary stage to facultative eusociality crosses the threshold to eusociality and occurs via exaptation[23] and is

---

[23] 'exaptation' (based on Wikipedia, 1st June 2016) denotes an evolutionary shift in the function of a trait; e.g., a trait can evolve because it served one particular function, but subsequently it may come to serve another. Koonin in [9] (*page 1014*) explains it slightly different: "The spandrel metaphor holds that many functionally important elements of biological organization did not evolve as specific devices to

characterised by the appearance of the allomaternal care task, synonymous with the appearance of worker behaviour. The shift from facultative to primitive eusociality takes place via phenotypic accommodation "in which the consistent repetition of caste-like behaviours across generations enables selection to act in a context of nongenetic mechanisms (ref)." ([138] *page 6*). The final shift from primitive to advanced eusociality takes place via genetic assimilation, mediated through the "constancy of phenotypic accommodation provides context for fixation of regulatory mechanisms or novel alleles via genetic assimilation (*refs*)." ([138] *page 6*). All primitive eusocial wasps "can express maternal behaviours, and maternal and allomaternal phenotypes can transition from one to the other ..." ([138] *page 5*), while advanced eusocial wasps have fixed maternal and allomaternal phenotypes. The nature of the selection in Hunt's model is a gradient; obviously, selection at the solitary stage operates at the individual level, while the selection has to operate at the colony level for advanced eusocial colonies, because of the phenotypically fixed cooperative castes. In between these two extreme stages, there will be a mix of selection levels – that is, multilevel selection.

Based on the comparison of the genomes of two bumblebee species (characterised by a lower level of sociality) with that of ants and honeybees, Libbrecht and Keller in their editorial briefly highlight two potential mechanisms for the making of eusociality: 1) based on binding of the juvenile hormone: "... JH signalling pathway plays a key role in regulating division of labour in eusocial Hymenoptera." ([121] *page 2*); and 2) based on microRNAs, since "... microRNAs are involved in the regulation of many aspects of the honeybee social life." ([121] *page 2*). Libbrecht and co-workers (in an earlier paper) summarize the literature like this: "In some species, caste determination stems mostly from a developmental switch controlled by environmental factors, whereas, in others, strong genetic effects can also influence the process of caste determination. These genetic influences range from plastic genotypes that are biased towards queen or worker development to a strictly genetic determination [*ref*]." ([142] *page 1*). The authors review the caste determination and division of roles for the various ant species, expressing that "Recent studies implicate epigenetic processes in caste determination during larval development." ([142] *page 1*), highlighting examples for various ant species. The authors summarise studies on the mechanisms behind the evolution of sociality by comparing genome characteristics between social and solitary insects. Several of these studies are still suggestive, but show that the epigenetic mechanisms in action in these species vary from being based on, for instance: 1) caste-specific gene expression and alternative splicing, implemented as diversity of transcription factor binding factors or reduced number of CpG islands; 2) hormonal signalling of environmental cues, implemented through type or number of receptors, through multiple genes coding for the specific hormone that regulate caste- and behaviour-specific expression; 3) differentiated lineages combined with hybridization at mating, where inter-lineage offspring develops into workers, while intra-specific offspring develop into queens; and 4) regulated through supergenes showing properties typical of sex chromosomes.

Cf. the note on perspectives in the introduction to this chapter, there may be details in Section 8.2.2 that in reality belong in the present section and will be moved when more details on the underlying mechanisms are known.

---

perform their current functions but rather are products of non-adaptive architectural constraints – much like spandrels...".

#### *8.3.2.1 Social Patterning and Behavioural Syndromes*

The fitness related to the social patterning and behavioural syndromes is a dynamically changing variable in evolution.

Behavioural syndromes (animal personalities) are consistent and/or correlated behaviours across two or more situations within a population, for instance some are consistently more aggressive or bolder than others ([143]). "Social insects can exhibit behavioural variation at multiple levels of organisation: between species, colonies, castes, individuals, and genetic lines (*ref*)." ([143] *page 49*). "Genetic studies show that behavioural types typically exhibit low to moderate heritability (…; *refs*), … often associated with neuroendocrine pathways, that explain some of the variation among behavioural types (*refs*)." ([143] *page 50*). Furthermore, the authors express that "Genetic effects on behavioural types are also typically moderated by individual experience. In particular, early experience can often have strong effects on the development of a later behavioural type and associated 'stress response' systems (*refs*)." ([143] *page 50*). This draws the attention toward potential epigenetic mechanisms as for the life-history effects.

Jandt and co-workers in their review provide a handful of examples of how an individual's behaviour can affect its fitness, in a context-dependent way ([143]):

- In a safe environment, selection might favour bold individuals for daring to do certain efforts to get food, but in dangerous environments cautious individuals might more likely survive
- Bolder individuals can be inappropriately bold when predators are present
- Aggressive individuals can be good at defending the nest (out-competing other colonies through interference competition), but a side-effect of such behaviour may be inappropriate aggressiveness toward mates and/or exhibition of poor parental care
- In some cases, behavioural differences might lead to task groups or behavioural specialisation, while inter-individual variation within such groups can still be observed
- In honey bees (*Apis Mellifera*), a threshold toward sucrose correlates with a handful of different behaviours, such as likelihood of collecting non-food rewards, likelihood of responding to the less-concentrated nectar at foraging, and the ability to learn
- Groups with a mixture of aggression types tend to have higher fitness than groups with only one type
- Bumble bee colonies vary in their cognitive behavioural type (learning speed) and foraging

Further, Jandt and co-workers in [143] say that:

- "The reproductive unit of social insects on which natural selection acts is the colony." ([143] *page 56*)
- Some social insects often change their role or their task repertoire expands over the course of their life, called temporal polyethism, which might be a type of maturation or ageing syndrome, while such switch in others it seems to happen at random; and often influenced by juvenile hormone levels
- A number of environmental factors can influence behavioural type, such as food type and availability and/or competition, temperature and social environment (e.g. interaction rates and communication signals)

- Studies where the researchers have removed individuals performing a certain task from a colony show that other individuals consequently change their task
- "The relationship between colony composition and emergent behaviour might be non-linear, i.e. just a few individuals could significantly change the behaviour of a group or population as a whole." ([143] *page 57*)
- Key individuals (referred to as catalysts, performers or organisers) can increase performance by better facilitation of information flow, by increasing the rate of activity of others, by motivating others to begin foraging, by communicating passionately, or they may increase cohesion within the colony – in honey bee colonies by vibrational signalling ([143])
- Several studies have indicated that feedback linking an individual's behavioural type and its energy reserves, condition or social rank may be a key mechanism for maintaining consistent differences in behavioural types

The review of Jeanson and Weidenmüller, [144], nicely supplements the review of Jandt and co-workers by focussing specifically on the various aspects in relation to inter-individual variability, cause and consequence. They emphasize that "Colonies of social insects however differ from other animals living in social groups in one important aspect; they represent functionally integrated and adaptive units. Selection acts both at the level of the individual and at the level of the colony … ." ([144] *page* 672).

Regarding how inter-individual differences arise, Jeanson and Weidenmüller say in [144]:

1) Inter-individual variability in behaviour is crucial for task allocation in terms of response-threshold mechanisms – that is, the individuals differ with respect to the stimulus level which enforces them to begin performing a given task; a minimal degree of intra-individual behaviour is however, required;
2) One way of diversifying behavioural responses is through genetic variability: a) polyandry (the queen mates with multiple males), for instance in honey bees empirical evidence indicates that workers differ in their sensitivity to task-associated stimuli as well as in task performance for a variety of activities: b) polygyny (a colony has multiple queens); c) recombination;
3) Another way of diversifying behavioural responses is through phenotypic plasticity: a) variation in environmental conditions, for instance, rearing temperature affect onset of foraging and a higher probability of dancing in honey bees; b) age-related (maturational) shift in task allocation (e.g. linked to nutritional and reproductive signalling pathways including insulin, juvenile hormone, vitellogenin and/or the expression of the foraging gene), where such shift is accompanied by "massive changes in the brain gene expression" ([144] *page 675*); c) nutritional state can modulate the response threshold to task-associated stimuli as well as sensitivity to external cues (e.g. individual hunger can regulate foraging behaviour in honey bees); d) "Experience may be one of the most important, omnipresent parameters modifying individual behaviour and driving behavioural differentiation among the workers in an insect colony." ([144] *page* 676) throughout life, and that "almost all behavioural responses are fine-tuned through learning" ([144] *page* 676), correlated with neuroanatomical changes – and even through passive exposure to stimuli such as those in the first days in the life of a social insect; e) social environment in terms of group size and feedback loops between nest mates or the opposite – being kept out of certain information loops;

Jeanson and Weidenmüller further reviews the effect of variability on the colony: Overall, it seems that inter-individual variability is beneficial to colonies, "mostly based on the existence of more efficient patterns of task allocation leading to improved task performance, a reduction of fluctuations in task-associated stimuli and a higher resilience in response to external perturbations." ([144] *page* 679). Additionally, the co-existence of various genotypes has turned out to reduce the impact of infection within colonies. Beyond this, they report that some studies show a positive correlation between genomic variability and colony performance, while others find no positive influence.

A further couple of reviews that are of relevance here, although a little more marginally in their entirety, are Oliveira's [145] and [146]. Oliveira explicitly states that "social competence, defined as the ability of an animal to optimize the expression of its social behaviour as a function of available social information, should be considered as a performance trait that impacts on the Darwinian fitness of the animal." ([145] *page 423*) and continues with "Social competence is based on behavioural plasticity, which, in turn, can be achieved by different neural mechanisms of plasticity, namely by rewiring or by biochemically switching nodes of a putative neural network underlying social behaviour." ([145] *page 423*). Further, the author suggests and provides substantial pieces of evidence for how such neural mechanisms of plasticity may operate ([145], [146]), and that monoamines, neuropeptides, and sex steroids may act as neuromodulators regulating context-appropriate behaviour. That the wiring of the brain is an anatomical consequence (and vice versa) of social behaviours was already discussed in Section 5.2.4, namely that altruism and reciprocal altruism are distinct behavioural actions implemented as different neural connections. A recent similar observation by Smith and co-workers from the Human Connectome Project ([147]) also links brain connectivity and behaviour, as they observe a difference in brain connectivity between positively and negatively thinking humans, showing that at least some behavioural patterns are explicitly wired within the brain. So, according to the statement above on the implementation of social behaviour being controlled by plasticity (which was not obvious from the paper of Hein and co-workers ([76]), and affecting the fitness in a social environment, the example and parts of the 'kin selection' section should probably be moved to or at least partially duplicated in the Extended Evolutionary Synthesis(!?) – time will show when evidence for other species than Oliveira's model species, fish, are repeated.

Furthermore, an example of cultural differences that affects the fitness of individuals: The survival criteria in historic times were related to the nature of the society of that era. For instance, agriculture and industrial production are based on hard physical work and low mental load. Today's Information Age in the Western cultures is almost a polar opposite with a shift in job profiles toward low physical and high mental load, like clerical type of jobs and for instance jobs related to the information technology. These are, for instance, skills related to the development, operation and maintenance of Information Technology applications in the widest sense are called for. Such jobs tend to be medium and university level educations and hence upper middle social class salaries, rather than the low skilled workers' salary in agriculture and industrial production. Consequently, people with less physical/biological fitness may thrive economically and socially in the Information Age as compared to previous eras. Thereby, the technology-induced change has also changed the fitness landscape, and consequently those who are now best fit within the societal structure and its need for specialist competences increase their share of the gene pool through providing better conditions for their offspring.

Finally, like for other animals, individuals with the most offspring also have a higher chance of promoting their own gene pool, which in some places/cultures and hence in some immigrant groups in the Western countries tend to be a characteristic of lower social classes. Also – as a result of the social patterns and career preferences – women (in the Western countries) tend to get fewer kids later, and fewer in the higher social classes than in lower social classes. So, in some cases, it may be the economically less fit that generates and thereby propagates the most progeny, which one would anticipate to be opposite to fitness. On the other hand, in a 'fitness' context, a confounding factor might be that the economically and socially fit mothers may still overall provide better support and competitiveness for their progeny, for instance in terms of better educations (i.e. the concept of 'parental investment').

In conclusion, all of the above indicates that fitness in a social and cultural context is complex.

### *8.3.3 Time and Timing*

Since Time and timing aspects are major issues in Function 5, this is explicitly included in the Unifying Theory model in Table 1, and there is one obvious theory that belongs here: Life History Theory. Further, there are a couple of other topics that shall also be briefly discussed in the following.

#### *8.3.3.1 Life-History Theory*

> Central to an evolutionary-developmental perspective is the concept of conditional adaptation: "evolved mechanisms that detect and respond to specific features of childhood environments, features that have proven reliable over evolutionary time in predicting the nature of the social and physical world into which children will mature, and entrain development pathways that reliably matched those features during a species' natural selective history...".
>
> Ellis et al. ([148] *page 94*).

Life history theory is a branch of evolutionary biology concerned with how and why organisms allocate resources (time and energy) to various competing activities over their life cycle; see for instance [149], [150], [151], [152]. "Life history theory attempts to explain life history traits in terms of adaptive trade-offs in distribution of resources to competing life functions: maintenance, growth, and reproduction (*refs*)." ([149] *page 26*). Note that conscious thought is generally believed to not play a role in the regulation of life history ([149], [150], [151], [152]), while such decisions are regulated by cascades of conditional developmental switches (West-Eberhardt cited in [149]; see also the elaborate discussion on this topic in the review [152]); each switch may be regulated by thresholds controlled through (epi)genetic settings and environmental conditions, and may hence act as discrete states.

Benton and co-workers say that "heterogeneity between individuals carries key information that should not be discarded in favour of finding a population average parameter." ([64] *page 1173*). This, however, does not (necessarily) rule out the general principles in the present perspective. The continued modelling according to the template information model with the three perspectives would fall apart if this was the case. The review of Benton and co-workers specifically is concerned with the importance of individual variability in understanding dynamics. They point out that "an individuals' response to a given environment depends on both the current environment and the past environment." ([64] *page 1174*), discussing the life-cycle's effect on population structure (distribution of ages and stages of individuals), plasticity, trade-offs and intergenerational effects. For instance, experiments with soil mites (*Sancassania Berlesei*) show that adults respond to food supply by increasing their fecundity, while juveniles respond by growing; so the distribution on age/stage of the individuals in the population in itself influence the population dynamics.

When one addresses the issue of development – that is, childhood and adolescence, Time and timing are built-in circumstances. There is a meticulous progression from the zygote to the fully developed phenotype, a process that is controlled through epigenetic switches and numerous factors. It is therefore, obvious that plasticity may be a function of time, involving both absolute time ('Time') and timing. Beldade and co-workers ([33] *page 1354*) express the timing issues in the sensitivities toward environmental cues as follows: "They also exist in relation to restricted time windows of the development during which the external environment can influence the outcome ..., development being quite robust outside these sensitive periods". Similarly, the time (duration) of a process may be affected, either uniformly extending or reducing the total duration of development or for a specific stage (affecting the 'tissue-by-stage-specific gene expression' [33]). That is, a combination of the Time and timing issues.

Ellis and co-workers in their introduction summarize the literature, saying "Epidemiological evidence … indicates that early-maturing girls, …, are at elevated risk for a variety of negative physical and mental health outcomes, including unhealthy weight gain, early initiation of substance use, early sexual initiation and pregnancy, emotional and behavioural problems, and mortality from cardiovascular disease and breast cancer (*refs*)." ([148] *page 85*), and "Beyond the effects of pubertal timing, pubertal tempo, …, has also been implicated in development of psychopathology and physical health problems." ([148] *page 85*). This seems convincing regarding the effect of pubertal maturation as a factor in evolution through its direct and indirect effects on fitness and reproduction.

Another specific example further makes certain that there are other behavioural patterns for humans that are influenced by early-life environmental conditions and which may therefore indirectly affect such individuals' reproductive capability: From ([1] *page 500*):

> "Communicative transactions between a phenotype and its environment (i.e. often parents) in terms of social interactions may even account for acquired pathological behavioural patterns and the accompanying anatomical changes found in persons. One such example is found with borderline personality disorders (BPD). A meta-analysis revealed a significantly smaller volume in both hippocampi and amygdala of BPD patients (compared to healthy controls), these brain areas playing a central/critical role in controlling emotional reactivity (*ref*). Dammann and co-workers (*ref*) report statistically significant increased DNA methylation patterns, and conclude *'Our data suggest that aberrant epigenetic regulation of neuropsychiatric genes may contribute to the pathogenesis of BPD.'* (*ref*). In this respect, the review (*ref*) reports that *'... both genetic and environmental variables appear to have the most profound and enduring effects when they exert their effects during early postnatal periods, times when the forebrain is undergoing exuberant experience-expectant dendritic and axonal growth;'* …." (end of citation from [1]).

This is further reviewed and analysed in three more recent reviews, confirming the early-life environmental effect on the development of personality disorders and shows that the issue may be more complex, but no doubt that there is a correlation between early-life insecurity, disorganised parent-child relationship and the personality disorder: [153], [154], [155].

Given the characteristics of BPD, it is anticipated that such behavioural trait has altered altruistic and cooperative behaviours of the affected individual as well as weakened his/her sexual/familial relationships, and hence that the disorder will influence evolution for the affected individual through its fitness within a social context.

The key units of analysis in life history theory are what is called 'life history traits', "characteristics that determine rates of reproduction and associated growth, aging, and parental investment (such as, size and number of offspring, amount of investment per offspring, age at sexual maturity, time to first reproduction, longevity)." ([149] *page 26*).

Belsky and co-workers in [150] report from a study that prospectively investigated the predictability of early sexual behaviour as a latent measure of an accelerated life history, based on early-life environmental harshness and unpredictability, which are assumed to be independent of one another. Harshness was measured in terms of 'income-to-needs' (assuming that limited economic resources challenge the coping capacities of families), while a measure of environmental unpredictability is based on a combination of paternal transitions, household moves and employment transitions. The conclusion of Belsky and co-workers' study was that "... low income-to-needs and greater environmental unpredictability in the first 4.5 years of life each uniquely predicted increased maternal depressive symptoms across the toddler and preschool years, which itself predicted less maternal sensitivity during the middle-school years and, thereby increased sexual activity in adolescence." ([150] *page 671*). They continue with "And even with these two pathways taken into account, greater environmental unpredictability, though not lower income-to-needs, forecast greater adolescent sexual activity in a direct, unmediated fashion." ([150] *page 671*). In short, they argue along the 'weather forecast' philosophy that "...developmental systems have been shaped by natural selection to respond adaptively to both "positive" and "negative" developmental contexts." ([150] *page 662*), and "... parenting provides young children with a "weather forecast" of sorts, alerting them to what they should "wear" —cognitively, emotionally, physiologically, and behaviourally—in order to succeed in the fundamental tasks of growing, mating, and eventually reproducing (*refs*)." ([150] *page 663*); simply because "...natural selection favors individuals able to "schedule" development and activities (i.e. allocate resources) in a manner that optimizes trade-offs over the life course and other varying ecological conditions." ([150] *page 663*).

Another similar, prospective study by Simpson and co-workers investigates the harshness and unpredictability from before birth as predictive measures of life-history outcomes in terms of age of sexual debut, number of sexual partners, multiple indicators of risk taking and criminality, distinguishing age 0-5 from age 6-16 ([151]). Because of the social characteristics of the recruiting area (as emphasised by the authors), this study may be a better representative of the intensity and duration of life stressors, compared to Belsky and co-workers' study. They strongly conclude that exposure to an unpredictable environment between age 0 and 5 years is an indicator for more sexual partners, and that it prospectively and independently predicts more lifetime sexual partners, higher levels of aggression, delinquency, and criminal behaviour, while neither harshness nor unpredictability are indicators for any of the three outcome measures in the 6-16 year group, – measured by the age of 23.

Brumbach and co-workers have other definitions of harshness ("general physical strain", i.e. "extrinsic mortality risk") and unpredictability ("the degree to which there is unpredictable variability in the outcomes of adaptively significant behaviour (*ref*)." ([149] *page 28*) (i.e. "unreliable home life" ([149] *page 32*)) than the above, but investigate the same effect measures. Moreover, according to [149], there are three types of life-history traits: biological, behavioural and cultural. This is the only study of this kind identified where the cultural aspects are explicitly included; however, with the present definition of culture above, religious belief as one facet of culture may be a strong predictor of sexual initiation and behaviour – and at some level a conscious decision and thereby contradicting the assumption of unconscious decision making. Nevertheless, such inconsistency between studies will probably not ruin the main idea of the life history theory or the studies referenced here, as they are all performed in western countries, and because the cultural aspect is more or less ignored in the measures of the independent variables in the above reported studies, while Brumbach and co-workers' own prospective study include some (culturally determined) life style aspects, such as sexual activity and attitude toward contraception. These authors detail their independent measures as follows: direct mortality risk (harshness) is operationalised as "self-reported exposure to violence from conspecifics", and unpredictability is operationalised by "frequent changes or ongoing inconsistency in several dimensions of childhood environment" ([149] *pages 32-33*); all in all they have an impressive list of measures. Their findings indicate that environmental unpredictability and harshness in adolescence have direct and indirect effects leading to a faster life history strategy[24] in young adulthood. They conclude overall: "that the harshness and unpredictability of the environment were linked to adolescent life history traits, that environmental unpredictability continued to have lasting direct effects in young adulthood, and that both harshness and unpredictability had indirect effects on young adult life history strategy through the adolescent life history factors." ([149] *page 45*), and that the effect does not come from general stress, but that the primary influence comes from "specifically harshness (mortality risk) and unpredictability (stochasticity) …" ([149] *page 46*). Rickard and co-workers [156] argue along a similar thought of harshness as having an external detrimental effect on the body's soma, a long-term effect caused by early circumstances on the body, and with the same end result as presented by others.

Further, Ellis and co-workers in their introduction summarize the literature with "…children displaying high BSC appear to have higher levels of mental and physical morbidities under harsh or stressful conditions but unusually low level within supportive and protective conditions (*refs*)."[25] ([148] *page 86*), which was confirmed by their study.

Still further, Hummel and co-workers in their systematic review, [157], find similar conclusions for adolescent substance use as a function of family functioning and pubertal timing, but in particular for girls.

Nettle and co-workers in [158] emphasise that the relationship between early-life events and adult life-history milestones are rather small effects for single early-life

---

[24] 'Life history strategy' is the development of a coherent, integrated suite of life-history traits in terms of an unending chain of decision nodes over the life course regarding resource allocation ([149]), and "The term "strategy" denotes an organism's realized phenotype among a set of possible alternatives. ([152]).

[25] BSC = Biological Sensitivity to Context

variables, but larger effect size when multiple early-life variables are combined into overall indices. In this respect, Davidowitz and co-workers ([159]) show that when selection on the life history traits pulls in the same direction then these act synergistically, and antagonistically when selection pulls in opposite directions. Furthermore, in this respect, Roff in his review – where he includes other species than humans – states 1) that "In a population or quantitative genetic context a trade-off is understood to be the result of either linkage disequilibrium or antagonistic pleiotropy, with antagonistic pleiotropy being the more commonly assumed mode of action for trade-offs between life-history traits." ([160] *page 120*); 2) that "Both artificial selection and experimental evolution have demonstrated the presence of genetically based trade-offs, consistent with the central assumption …" ([160] *page 123*); and 3) that "…there might be combinations of parameters at which fitness is not maximized but from which any small deviation reduces fitness, thereby 'trapping' the population on a suboptimal fitness 'peak'." ([160] *page 123*), explaining how trade-offs may lead to the various paths of adaptation to external stimuli. Ellis in [152] reviews the role of a particular hormonal system that enables a switch-controlled and condition-sensitive system; such a mechanism in combination with the other suggestions would nicely fit and explain the system implementing the adaptive principle in life-history strategies.

Sonuga-Barke in a reward commentary expresses that both twin and adoption studies confirm the importance of genetic factors in conduct disorders (abbrev. "CD") and temporal discounting (discounting of utility assigned to future rewards depending on the anticipated delay), "At the cognitive level, exposures to the uncertainty associated with psychosocial adversity during development may create a deep-seated mindset in the individual with CD in which delay and uncertainty are inseparably linked during the coding of future outcomes. This could manifest itself as a cognitive bias in decision processes in which preference for future rewards is reduced, …" ([161] *page 22*); no doubt this may influence altruistic behaviours and cooperation. The author hypothesises a combination of three named circuit links in the brain that may be compromised and which may explain the temporal discount in conduct disorders: i) the links known to play a role in prospection about future outcomes, goal setting and plan implementation; ii) the links known to play a specialised role in deliberative processes; and iii) the link to existing findings described earlier.

The conclusion is that early life conditions have an effect on the biological development as well as on the behavioural patterns later in life. That this is a statistically significant correlation fortunately only makes it a question of probabilities, or there would have been no hope for the Romanian institutionalized children from the era of Ceauşescu: A study on English and Romanian adoptees concludes that pure psychological deprivation (in the absence of sub-nutrition) had a profound effect on psychological functioning in the form of deprivation specific patterns ([162]), and another study that 20% of adopted children from Romania suffered from Posttraumatic stress disorder ([163], see also [164]). A third study on Romanian adoptees in the US states that their study "…illustrates that a history of institutionalization had minimal long-term adverse effects on a child's behavioural health." ([165] *abstract*); but unfortunately, the present author cannot access the article as a whole.

Still, a prospective study by Tung and co-workers ([166]) demonstrates for a population of individually recognized baboons observed continuously during the period of 1983-2013 that "Specifically, females who experience more cumulative early adversity have significantly shorter adult lifespans–on the order of years–which translates into fewer surviving offspring and lower lifetime reproductive success." ([166] *page 2*).

Since baboons reproduce throughout life and rarely experience a post-reproductive period, such shortening of the lifespan (a reduction of about 10 years in the median lifetime) has a highly significant influence on fitness in terms of reproduction of such individuals. The factors observed were: 1) drought in the first year; 2) population density and hence competition; 3) maternal dominance rank; 4) maternal affiliative social connectedness; 5) maternal loss before age 4 years; 6) competing younger siblings. Some of these factors, according to the authors, are mediated through resource limitation, and correlated with adult social relations and maternal investment (for factor (6)).

In summary, the remark by Brumbach and co-workers in [149] that life history is regulated through a cascade of conditional developmental switches makes it plausible that Time and timing – as all of the above studies document – is a significant factor in evolutionary biology.

#### *8.3.3.2 Miscellaneous on Time and Timing*

Obvious additional candidate suggestions for 'Time & timing' aspects in relation to the Extended Evolutionary Synthesis would be:

- The wash-out phenomena, where the plastic adaptation evades either in the affected phenotype or gradually over a couple of successive generations of progeny; see the discussion in Burggren's review ([167]). The wash-out phenomenon (and a corresponding wash-in phenomenon of plastic changes) clearly shows that 'Time' is an issue in evolution. Burggren stresses that literature on this topic is still sparse, while highlighting the existing evidence. For instance, "In thermoregulating *Bombus terristris*, repeated performance of the task of fanning leads to a decrease in the corresponding temperature response thresholds (*ref*); a 'reinforcement' process that seems to be reversible when the task is not performed for a while (*ref*)." ([144] *page* 677).
- Speed of adaptation, or similarly a delay in its implementation; the latter corresponds to the 'wash-in' phenomenon mentioned in the previous bullet.
- An effect of plasticity comes into play in time because the time to reach maturity or longevity can lead to biologically important dynamical lags being context-dependent rather than of a fixed time-step [64].
- "Long-term fitness reflects the performance over many generations (…), and there is generally a single strategy that maximizes this long-term measure of evolutionary success (…). Therefore, a global monomorphism is to be expected. However, the resulting genotype will often be a diversifying 'bet-hedging' strategy (*refs*), …" resulting in a stochastic distribution of two or more phenotypes over a couple of generations ([79] *page 3965*).

The transgenerational epigenetics compares to a heritable epigenetic setting and its advantage is that it may allow for time to acquire permanent genomic changes for coping with a given environmental condition through the mechanism of fixation. Epigenetic settings are not heritable in the same way as genomic changes. However, transgenerational epigenetic settings have clearly been demonstrated in terms of pathological conditions that are transferred across generations either in the maternal or the paternal line; see chapters throughout [38], and also reviews like [168].

Another timing aspect is the menopause. Evolutionary theories based on natural selection predict that menopause should not occur because there would be no selection for survival after the cessation of reproductive ability, and yet there is more of interest in this respect. A couple of cohort studies on breeding and survival patterns were briefly discussed in [1], suggesting a social variant of altruism that is correlated with the grandmother's geographical availability to support in bringing up her grandchildren and thereby favouring the fitness of her family. This was demonstrated valid for the time period of the cohort studies (1702-1823 for the Finnish cohort and 1850-1879 for the Canadian cohort), and supports Darwin's theory of survival of the fittest also in a Time and timing perspective. If the mechanism was valid then, it will be valid now, however, today's condition for living (in the Western culture) implies that both parents have full-time jobs, and hence this condition adds a career factor into the timing 'equation'. Therefore, grandmothers now often have their own professional career and hence may not be available to support the up-bringing of their grandchildren other than babysitting now and then as well as supporting knowledge-wise. From a fitness perspective, this merely implies that the role of grandmothers has changed in practice and therefore has become a less visible factor in evolution.

Also killer whales are able to live for a significant time span after their menopause, actually up to 30 years. A study on female killer whales shows that they support their progeny and kin through acting as 'repositories' of accumulated ecological knowledge, knowledge of foraging ground and how to handle scenarios of low food resources. They become the leaders of their group, and this is particularly prominent in situations with scarce food ([169]). A similar pattern of leadership is also observed in relation to collective decision-making (see Section 7.2). The same is the case in some human cultures, where the grey-haired generation is generally appreciated and respected *per se*.

### *8.3.4 Fidelity*

Fidelity is concerned with how to ensure the quality of the phenotype at any given time – that is, it includes both quality and risk management. Also in an evolutionary context, risk management and quality management are two sides of one coin in that the latter continuously seeks to optimise quality while the former strives to optimise the balance between advantageous changes and detrimental changes, and in the scenario of evolution also taking into account that the latter may one day become advantageous and therefore at times it may be worth taking chances.

A couple of points in relation to the principles behind the Extended Evolutionary Synthesis theory come to mind in this respect:

1) Plasticity confers robustness to the development of a phenotype ensuring adaptation to present environmental conditions and thereby increasing the immediate chances of survival;
2) "... because phenotypic plasticity can shield genetic variation from natural selection, it can presumably promote the accumulation of cryptic variation (i.e. genetic variation that does not result in phenotype variation)." ([33] *page 1353*; see also [47]). Cryptic variation may in some cases be an advantage as it constitutes a buffer of latent changes and hence may be invoked in a situation urging for such change in a trait;

3) Plasticity can accelerate adaptive evolution (see e.g. [33], [170] and many more), thereby optimising the phenotype traits to the environmental conditions;
4) Another advantage of plasticity is the cleaning of the 'board' from one generation to the next, as opposed to DNA-driven evolution. Therefore, if an environmental change persists and every individual in a population adapt by means of plasticity then there will be a large population in which changes in the genome may accumulate to fix the change into a new and potentially stable condition; see Section 6.3;
5) If an environmental change persists, the cleaning between generations will enable new variants of the advantageous plasticity to thrive even further while the old environmental condition would not have been able to test (i.e. select) such new and potentially more effective capability pattern in real practice – that is, prior to a selection pressure;
6) Epigenetic inheritance can both facilitate and retard plasticity, depending on the details of the inheritance, interactions with the genetic and environmental factors, as well as environmental variation ([35]). The wash-out phenomenon previously mentioned further shows that transgenerational epigenetic changes may be reversible, which is an advantage at transitory external conditions;
7) Variable transmission fidelity is expected if there is a reversal of induced epigenetic states in the absence of the inducing factor ([35]);
8) Plasticity may imitate speciation in that in a situation of sparse resources some individuals in a population may develop a preference for a different type of food and thereby reduce competition for food; see for instance ([33]).

Facultative parthenogenesis (asexual reproduction, in which the development of an embryo occurs without fertilization) has been reported for a wild critically endangered vertebrate, the smalltooth sawfish, which normally reproduces sexually; see details in [171]. Previously, parthenogenesis is only found in sexually reproducing vertebrates without genomic imprinting. It might be that such change from sexual reproduction to parthenogenesis is risk management controlled by plasticity – that is, through epigenetic mechanisms as a response to combat extinction; and therefore, it belongs here in the model. The biological advantage of facultative parthenogenesis is that extinction is delayed, as the female individuals survive (in terms of their cloned progeny) for some time and thereby they are provided with a chance of finding a male to mate with. Naturally this will have major implications for the genotype and phenotype of the ('cloned') progeny because of the extreme relatedness.

# 9 Evolution of Evolvability, Survival of the Fittest Evolutionary Mechanisms

**Abstract.** Function 6, evolution (see Table 2 and Table 3 in Chapter 14), is concerned with evolvability, "the capacity of a system to evolve", which includes both the amount of variation and the nature of such variation. Every resource is susceptible to the system's evolutionary mechanisms, directly and/or indirectly, and hence so are the mechanisms creating evolution. The important here is that the system ensures sustainability as well as (re)generation of the phenotype – or alternatively: its progeny – after a given evolutionary change. The principle of this function is 'survival of the fittest evolutionary capability', and the mechanism is 'natural selection among evolutionary mechanisms', explained for the three perspectives: Standard Evolutionary Theory, Inevitable Evolution Theory and Extended Evolutionary Synthesis.

**Keywords:** Natural selection, survival of the fittest, Standard Evolutionary Theory, Extended Evolutionary Synthesis, Inevitable Evolution Theory, constructive neutral evolution, inevitable evolution, evolvability, homeorhesis, horizontal gene transfer, selfish genes

In Kirschner terms, 'evolution of the evolvability' ([7] *page 1*), where 'evolvability' means "the capacity of a system to evolve", includes both the amount of variation and the nature of such variation. Kirschner points at 'evolvability' as a constituent principle in evolution and elaborates on the concept of evolvability in relation to Darwin's theory of natural selection and the genetic mechanisms for generating variation.

**Principle**: Survival of the fittest evolutionary capability
**Mechanism**: Natural selection[26] among evolutionary mechanisms
**Emergent property:** Selected phenotypes exhibiting sustainable mechanisms for establishing the variance necessary for continued future evolution

Every resource is susceptible to the system's evolutionary mechanisms, directly and/or indirectly, and hence so are the mechanisms creating evolution. The important here is that the system ensures sustainability as well as (re)generation of the phenotype – or alternatively: its progeny – after a given evolutionary change.

Natural selection at the level of evolutionary mechanisms is meant to give room for other/new principles and mechanisms.

[26] Note the definition of 'natural selection' in Section 1.2.

| **Inevitable Evolution Theory (IET)** | **Standard Evolutionary Theory (SET)** | **Extended Evolutionary Synthesis (EES)** |
|---|---|---|
| Constituent processes in evolvability:<br>• 'Inevitable evolution' as a result of stochastic variation | Acquired processes in evolvability:<br>• Evolution of 'evolvability' in a complex space of combinatorial opportunities<br>o Population genetics<br>o Horizontal gene transfer<br>o Selfish genes | Adaptive processes in evolvability:<br>• Homeorhesis<br>o Population genetics<br>o Constructive neutral evolution<br>o Horizontal gene transfer<br>o Selfish genes |

### *9.1 Inevitable Evolution, 'Inevitable Evolution Theory'*

Witting in his review, [22], suggests an evolutionary theory 'inevitable evolution' for which evolution is driven by a deterministic selection pressure acting on the energetic state of the organisms together with the density-dependent competition between interacting individuals in natural populations. He is not the first and only to suggest something similar to this theory, but so far this theory has not received much attention in the literature on evolution. At this position in the Unifying Theory, it is the principle of the evolvability that is addressed – that is, evolvability at the perspective of IET as inevitability of evolution.

The basic foundation for evolution within this perspective comprises the laws of physics, including quantum mechanics and thereby thermodynamics. One driver of the processes looked at for this perspective is variation in energetic states. In quantum mechanics, a state is not defined deterministically, but probabilistic. Similarly, fitness is perceived as a somewhat probabilistic measure rather than an absolute measure. Most phenotypes are multi-trait organisms and therefore, the fitness landscape is multi-dimensional. An analogy merely in 3D would be a landscape of mountains, valleys, rivers, passes and plateaus. Further, – to stay in the analogy – the mountains are not necessarily smooth and single-peaked, but may be multi-peaked and have rugged shapes, plateaus and walls, as well as valleys that may be deep or shallow, narrow or wide, or there may be a sea of neutrality. In a flat landscape of energetic states, stochastic variation may in a long term turn out being a determining factor moving evolution far and wide, while when being on a steep cliff stochastic variation may instantly push you off the cliff or across a small pass to a different energetic state, but not likely uphill. However, a fitness landscape has potentially many more dimensions than three, for instance, it might have one 'dimension' per evolutionary trait, and potentially inter-dependent in complex ways.

Solé and co-workers in [29] discuss whether self-organised criticality could be the mechanism behind major extinctions, and say that "In an ecological context, the addition of new species would place ecosystems close to a critical state, where the collective and not the individual species would be the relevant object in the long run." ([29] *page 158*). In contrast, Loreau and de Mazancourt begins their article by summarizing "There is mounting evidence that biodiversity increases the stability of ecosystem processes in changing environments, …" ([172] *page 106*), but also say that the mechanism beneath is poorly understood, and that the asynchrony of species' responses to environmental fluctuations may be a key element in the stabilisation. It may be that the two references are both right within each a set of circumscribed conditions, namely that the one is generally correct in realistic environments with natural fluctuations, but that there may be a tipping point where even a single new species makes the ecological system reach the critical point.

Witting proposes that "biotic evolution is driven by a universal natural selection where the long-term evolution of fitness-related traits is determined mainly by deterministic selection, while contingency is important predominantly for neutral traits." ([22] *page 259*). He says that selection by mechanisms involving energetic state and density-dependent competitive interactions is able to explain the huge transitions from simple asexually reproducing self-replicators to somatic, diploid organisms with sexual reproduction, senescence, and that it can explain even the evolution of eusocial colonial systems.

Witting [22] suggests that it seems to be traits linked to fitness that may evolve by deterministic natural selection, while the neutral traits are more susceptible to 'opportunities and chance' – that is, stochastic selection or natural selection by drift and diffusion.

Another argument that Witting mentions is parallel evolutionary trajectories in repeated experiments on fast-replicating organisms like E.Coli. Still another of his arguments in favour of Inevitable Evolution is that given organisms at sufficiently high levels of organisation develop several morphological traits that are obviously advantageous, like mobility and eyes, large body mass, etc. Witting's argument is that "…the widespread occurrence of convergent evolution, where phenotypic traits like eyes have evolved independently several times on Earth, indicates that there are deterministic-like selection pressures in wide ranges of organisms." ([22] *page 262*); vision is obviously an advantage selected for independently several times. Eusociality has evolved at least ten times (see [121]); envenomed structures in cartilaginous and bony fish have evolved independently 18 times (see [173]); and also lactose tolerance in humans has evolved several times simply concluded from the mutations behind the intolerance in various geographical areas; see for instance the review [108]. Recently, still another example of convergent evolution was reported by Foote and co-workers in relation to the genomes of marine mammals (killer whale, walrus and manatee) ([174]); they conclude in this respect that "Our data therefore indicate that although convergent phenotypic evolution can result from convergent molecular evolution, these cases are rare, and evolution more frequently makes use of different molecular pathways to reach the same phenotypic outcome." ([174] *page 274*).

Witting mentions other similar convergent evolutionary traits, such as body mass and brain size, of which at least some aspects are addressed in some of the references in Section 8.3. Alternative/competing mechanisms for controlling adult body size/weight have been demonstrated; the commentary of [175] suggests that there are (at least) two levels of control: i) nutrition-sensitive and mediated by insulin/TOR signalling, and is triggered at the minimum viable weight; and ii) the juvenile hormone overriding the insulin/TOR-mediated mechanism.

Therefore, – even if Witting mentions convergent evolution as an argument in favour of his evolutionary theory, at least all of these examples (including that of vision), are based on mutational and regulational evolutionary mechanisms and hence belongs elsewhere in the Unifying Theory of Evolution. So, Witting's principle of inevitable evolution must be based on all previous functions in the Unifying Theory of Evolution, and according to the template model this is OK, but it would have been nice with examples that are purely based on energy-states, stochasticity, and the similar. Self-organised criticality and punctuated equilibrium is one such example.

Doebeli and Ispolatov in their introduction summarise the highly complex nature of evolution for more complicated phenotype spaces containing scalar phenotypes of each of a number of coevolving populations: "Frequency dependence generates an evolutionary feedback loop, because selection pressures, which cause evolutionary change, change themselves as a population's phenotype distribution evolves." ([176] *page 1365*). Moreover, "… the frequency dependence can generate complicated evolutionary dynamics. For example, co-evolution of scalar traits in predator and prey populations can lead to arms races in the form of cyclic dynamics in phenotype space (*refs*), and coevolution of scalar traits in a three-species food chain can generate chaotic dynamics in phenotype space (*ref*)." ([176] *page 1366*). So, chaotic dynamics of at least long-term evolutionary trajectories may be relevant in the perspective of the Inevitable Evolution Theory.

Since this perspective is based on Laws of Physics, we should not expect evolution of evolvability in this perspective.

### *9.2 Evolution of Evolvability, 'Standard Evolutionary Theory'*

The basic mechanism in this perspective of evolution comprises opportunistic, non-directional perpetual change, such as mutations, replication errors, recombination (i.e. intrinsic causes from mitosis and meiosis related processes) and mutagens.

Already in year 2000, Tautz concludes in his review that "it is known that almost a third of all coding regions in an organism may show a fast evolutionary divergence, which could be a major reservoir for generating evolutionary novelties [*refs*]." ([52] *page 578*).

According to Koonin and Wolf ([24] *page 7*), few evolutionary biologist seems to believe "that evolvability is not selectable but is simply maintained at a sufficient level by inevitable errors at all levels of biological information processing.". As there are evolutionary mechanisms then even the inevitable errors from the information processing would be enough to change evolution itself – in a time perspective of millions of years and a corresponding number of generations. After all, the presently known evolutionary mechanisms cannot have been present in the first life forms, but must have evolved as a function of time after the appearance of the first proto-life based on RNA, and probably not all at the same time. Denying that evolutionary mechanisms can evolve would be to say that it all happened by coincidence the first time and will not happen again.

Evolvability is itself under selection as any other phenotypic trait and hence will change as a function of time and environmental conditions, because:

1) As Kirschner argues, "It is simply a byproduct of the evolution of physiological adaptability." ([7] *page 3*), and evolution applies the mechanisms that are anyway used by the body to create variability;

and/or,

2) Enhancing the evolvability may be achieved by increasing the capacity to tolerate non-lethal phenotypic changes exposed to selection, thereby exploiting the mechanisms for risk management discussed in Section 8.2.4. An example of this is the multiplication of genes and the introduction of polyploidy.

Further, Kirschner says "So I think many of the processes that people work on in biology with features that are puzzling and seem inexplicable will turn out to have explanations in terms of evolvability." ([7] *page 6*). For instance, polyploidy will enhance lineages that have a capacity to generate effective phenotypic variation, accumulating variation and keeping them 'hidden' until relevant and then refining such functionality to perfection. Polyploidy requires a set of mechanisms to work, like epigenetic erasure, imprinting, polarisation of the cell, etc. So, when did eukaryotes develop ploidy? This feature could be a distinguishing characteristic of lineages with extraordinary evolvability. Polyploidy is an evolutionary milestone that by means of recombination enabled a giant leap in the potential for variation, including positioning, as well as leading to the next milestone, evolution of sex and the meiosis.

Another example to Kirschner's statement right above is horizontal gene transfer, which has a dedicated section in the below.

Enhancing the capacity to tolerate non-lethal phenotypic changes will increase the capacity for accumulating more genetic variations.

Draghi and Wagner in [177] outline some theoretical controversies on whether evolvability can itself evolve by natural selection, while counter arguing and rejecting them all: 1) benefits to evolvability lie in the future and therefore will not affect fitness here and now; 2) selection for evolvability requires group selection; 3) an evolvable genotype does not survive its own success; 4) recombination will quickly dissociate an allele that improves variability from any positively selected variants it helps to create. Their own study "shows that mutant genotypes with higher evolvability are more likely to increase to fixation." and "that populations of highly evolvable genotypes are much less likely to be invaded by mutants with lower evolvability, and that this dynamic primarily shapes evolvability." ([177] *page 301*). The authors suggest that "Any trait that facilitates adaptive change in another aspect of the genotype is consequently insulated from the need to change and is under stabilising selection: this conservation may be a very general mechanism of selection on evolvability." ([177] *page 311*).

The highly complex combinatorial space of more or less independent potential in terms of existing genetic properties provides a testbed for evolution. And because every resource within the system is itself susceptible to the system's evolutionary mechanisms, evolution of evolvability is in itself merely specific and advantageous, successive changes on the path of natural variation.

This being said, there are also voices saying "that relatively simple, non-selective models might be sufficient to form the framework of a general evolutionary theory with respect to which purifying selection would provide boundary conditions (constraints) whereas positive, Darwinian selection (adaptation) would manifest itself as a quantitatively modest, even if functionally crucial modulator of the evolutionary process." ([9] *page 1025*).

In summary, "Darwinian evolution that is based on negative and positive selection acting on random mutations as well as genetic drift (Wrightian evolution) are intrinsic features of replicator systems, hence, in operation since the origin of the first replicators (that is, effectively, the origin of life) [*ref*]" ([78] *page 5*).

### *9.2.1 Population Genetics*

The genetic profiles of and within a population of individuals naturally affect evolution, as the fitness will vary among individuals and hence also populations. In a short-term perspective there may be an immediate effect, while the long-term effect depends on the ability of the traits in question either to be neutral (phenotypically expressed but without impact on fitness in the given context) and/or cryptic (phenotypically hidden or repressed).

The notable review of Romero and co-workers, [51], has a focus on the methodological pitfalls in comparing gene expression both within and between species, thereby showing us the caution needed at interpreting such publications. And as they say, "A major objective of evolutionary genetics is to provide a mechanistic account of the genetic basis for interspecies variation. The goal is to identify the genetic changes and molecular mechanisms that underlie phenotype diversity, as well as understand the evolutionary pressures under which phenotypic diversity evolves." ([51] *page 505*). The genetic profile with its potential for change is the basin for evolvability to evolve. That is why this topic is relevant in this place of the Unifying Theory of Evolution and in the parallel description for the Extended Evolutionary Synthesis perspective.

Romero and co-workers summarise a series of findings regarding evolution of gene expression – that is, one aspect of evolution in general, which implicitly embrace aspects of evolvability. Since the present unifying theory discriminates what takes place purely on the DNA sequence of the genome from the regulatory mechanisms of epigenetics, the present author has separated pieces of information from Romero and Co-workers' ([51]) into those belonging under this perspective and/or belonging under the Extended Evolutionary Synthesis (see Section 9.3.1). It is not always easy to discriminate when the underlying mechanisms are not fully explained, so, in case of uncertainty in this respect, the observation is kept in this place; however, a substantial fraction is expected to belong under the EES perspective since gene expression in the cellular metabolism is regulated epigenetically.

- "… expression patterns of many genes show remarkable conservation, suggesting a strong genetic component in their regulation as well as the action of stabilizing selection over hundreds of millions of years." ([51] *page 508*);
- Studies suggest that "evolutionary turnover of transcription factor binding sites is rapid and that, on a genome-wide scale, most binding locations may not be conserved even across closely related species" ([51] *page 510*); however, with Romero and co-authors expressing a (minor) caveat on the methodology of these studies;
- "Changes in *cis* elements appear to be more commonly responsible for interspecies differences in gene expression patterns than changes in *trans*, as shown in yeast and flies[*refs*]." ([51] *page 514*); however, note that the authors define '*trans*-regulatory elements' as "Regulatory elements that can affect the transcription rates of both alleles of a gene [examples include transcription factors and small regulatory RNAs]. By contrast, *cis*-regulatory elements have an allele-specific effect." ([51] *page 514*);
- "… several lines of evidence implicate chromatin state as an important player in the evolution of gene expression." ([51] *page 514*);
- "Recent experimental results in model systems[*refs*] are also resurrecting the classical idea that transposable elements, containing pre-existing transcription

factor binding sites, could insert in the vicinity of regulatory loci and could serve as a source of novel regulatory elements*ref*." ([51] *page 514*).

Even if these observations have been sorted, there may nevertheless be more that when digging into the causal mechanisms turn out to belong under the Extended Evolutionary Evolution perspective.

In summary, evolvability – even when strictly limited to population genetic considerations – is a function of changes within a complex space of combinatorial opportunities: given the many opportunities and combinations hereof, some inevitably and sooner or later will affect the mechanisms generating the source material for evolution and thereby promote evolvability. After all, one can follow the line back to the earliest life forms on Earth, where epigenetics did not exist, and compare with the evolutionary refinements shown in terms of the puzzle of theories within the Unifying Theory of Evolution, – that has to have evolved through the capability of evolvability in the perspective of the Standard Evolutionary Theory.

### *9.2.2 Horizontal Gene Transfer and Selfish Genes*

"Strikingly, in mammals sequences derived from mobile elements or and endogenous viruses account for at least 50% of the genome whereas in plants this fraction can reach 90% (*refs*)." ([24] *page 8*). Multiple mechanisms for such horizontal gene transfer – abbreviated 'HGT' – (exchange of genetic material between species) have been described, such as plasmid exchange, transduction[27] and transformation[28] ([24]). That this has contributed to who we are today may be illustrated by a conclusion referenced in [9] (*page 1015*) "that not only chloroplasts but also the mitochondria evolved from endosymbiont bacteria", namely from a particular group of α-proteobacteria.

Most of the prokaryotes do not engage in regular sex, and instead exchange genes with other microbes by means of HGT. HGT is extensive, pervades the entire prokaryote world and is the source of gene gain from bacterial and archaeal genomes (thus, naturally cohabitation is necessary), while also the opposite, loss of genes, is at least as prominent as the gain via HGT, and are the mechanisms that has literally shaped bacterial and archaeal genomes ([24]), so constructing a 'tree of life' for the prokaryote world does not give as much meaning as for eukaryotes.

One mechanism that appears to have specifically evolved to generate variation in prokaryotes is mediated by means of Diversity Generating Retroelements (DGR). In bacteriophages these retroelements "generate diversity in cell attachment surface proteins via reverse transcription-mediated mutagenesis, resulting in host tropism switching (*refs*), and in bacteria themselves where they produce receptor variation leading to bacteriophage resistance (*ref*)." ([24] *page 7*). As for the CRISPR-Cas mechanism outlined in Section 9.3.1, this mimics the animal immune system, except that the variation by this bacterial mechanism is inheritable, because the variation is included into the genome.

Recently, called Gene Transfer Agents (GTAs) were discovered (cf. Koonin and Wolf) another HGT mechanism, which is concerned with a distinct type of defective bacteriophages that package in the capsid rather than the phage genome. The GTAs

---

[27] Transduction is the transfer of genetic markers by bacteriophages ([24] *page 4*).

[28] Transformation is the acquisition of new traits via import of DNA from the environment and integration of the imported molecules into the bacterial genome ([24] *page 4*).

have been shown to infect and transfer their genetic content (from bacteria and archaea) to a broad range of other cohabitating prokaryotes ([24]).

Koonin and Wolf also express that one way to avoid the accumulation of deleterious mutations leading to a gradual decline in fitness is to enhance recombination via sexual reproduction (in the form of meiosis or bacterial conjugation) or HGT ([24]) – that is, the same mechanisms that enhance variation by horizontal gene transfer may also be the safety belt.

Another safety belt mechanism against the effect of viral infection – considered a kind of bet-hedging strategy – is the altruistic suicide mechanism that bacteria and archaea commit "using the toxin-antitoxin or abortive infection defence systems (*refs*)." ([24] *page 8*), thereby saving kin from the viral parasitic infection.

Koonin and Wolf express that "Recent, detailed studies indicate that at least in tight microbial communities, such as for instance the human gut microbiota, gene exchange is constant and rampant." ([24] *page 4*). Scary to even think of, when there might be a possibility of exchange also between the gut biota and its host.

In conclusion, using the words of Koonin and Wolf, the DGRs as well as the GTAs (and hence the HGT mechanisms) are "undeniably cases of evolution of evolvability" ([24] *page 10*). "We now realize that evolution of life is to a large extent shaped by the interaction (arms races but also cooperation) between genetic parasites (viruses and other selfish elements) and their cellular hosts." ([24] *page 10*).

### *9.3 Homeorhesis, 'Extended Evolutionary Synthesis'*

Homeorhesis (~momentum) for a system is its impetus to return to a trajectory of developments while adapting to its environmental and invironmental[29] conditions (as opposed to systems seeking to return to a stable state, called homeostasis).

Darwin insisted that evolution comes as a result of many small steps, while Kirschner ([7] *page 2*) says that "So Darwinian evolution is clearly a good mechanism for improving things – but it is not necessarily a good mechanism for generating novelty.". Following this, Kirschner argues that "The organism has the capacity to do a lot of different things physiologically…" and "along with ways of regulating them with feedbacks and mechanisms that constrain them so they are non-lethal…" ([7] *page 2*), and that such novelty after all (under the microscope) isn't that novel. Or, as Koonin says "…duplications … is a virtual death knell for Darwinian gradualism" ([9] *page 1022*). Moreover, looking back on the evolutionary mechanisms outlined in all of the above, it is not accurate to say that evolution comes as a result of only small steps; it is likely – again – a 'both/and' situation, a combination of small steps and major steps in between each other and which all fuel variation to the evolutionary process. Which is the most common is another question.

In [1] we posed the rhetoric question "Why is there an evolutionary principle like 'Survival of the fittest'?" or rather "how did it come into being?" From the perspective of the template information model, this is seen as a consequence of homeorhesis: constantly changing resources being released as progeny and re-introduced into the system as re-newed possibilities that are then further processed in the system in cycles of traversals through the seven functions. Random as well as non-random changes naturally and constantly occur, thereby creating a pool of variation (non-lethal, because the

[29] 'invironment' = inner environment ([2])

lethal ones are deselected if not otherwise compensated for). In support of this view, Ekstig concludes "The suggested model implies that complexity is cumulatively increasing, giving evolution a direction, an arrow of time, thus also implying that the latest emerging species will be the one with the highest level of complexity." ([126] *page 175*); however, he must refer to a given branch in the tree of life regarding his latter statement; otherwise the statement will not be valid. Ekstig references McShea and Brandon for "claiming that there is an even more fundamental biological law than natural selection: namely, the tendency for diversity and complexity to increase in evolutionary systems." ([126] *page 179*). It is the underlying mechanism that establishes whether such thinking belongs here in the Unifying Theory of Evolution, but so far, the suggestion is that this is Neo-Lamarckian thinking.

In the template information model, the system (in Function 7) releases progeny as renewed possibilities as a result of successfully finishing the sequence of functions. Consequently, such progeny – simply because every resource within the system is susceptible to the system's evolutionary mechanisms, directly or indirectly, and because evolutionary changes are not all mutually exclusive, will hold a number of changes – advantageous or detrimental. According to the principle 'Survival of the fittest', there will be a natural selection among the progeny resources, favouring those best fit for the current state of the system and its environment. However, the trajectory of homeorhetic development is unpredictable at large, although the non-random elements of change may give some indications of direction; see [3]. Evolution is itself changeable as a function of changing environmental conditions and the derived invironmental states. The pull of the homeorhetic development is to continue adapting to given conditions at any point in time. Therefore, we tend to disagree with the emotional expression in [24] (*page 1*) that "the cornerstone of Lamarck's worldview was the purported intrinsic drive of evolving organisms towards "perfection", a patently non-scientific irrational idea.", except that the notion' of 'perfection' should be replaced with 'the system's purpose' (i.e. specificity and directionality in terms of alignment with a purpose equals some aspects of perfection), where the system is any individual biological organism.

Moczek and co-workers say that "In many ways, evolution by genetic accommodation provides a shift in emphasis, rather than a radically new view of adaptive evolution." ([55] *page 2707*). Plasticity covers unexpressed genetic variation, and thus "...being unexpressed under a subset of conditions allows cryptic genetic variation that is neutral or even deleterious in some environments to persist in a population, analogous to models for recessive alleles." ([55] *page 2707*). Consequently, the environment through plasticity plays a formative role besides the role of selection. But when looking at this in a holistic view, it is not genetic accommodation and plasticity alone that have the credit; - there will always be an interplay with the other two perspectives in the Unifying Theory of Evolution somehow along the progress.

Koonin and Wolf in [78] mention repair mechanisms, evolvability mechanisms, stress-induced mutagenesis and plasticity mechanisms as candidate Neo-Lamarckian evolutionary mechanisms; but this might include also horizontal gene transfer, certain prions and more, potentially including reverse transcription (see Reviewer 2 in [78]). For instance, "Evolution of life forms of increasing complexity was enabled by increasing replication fidelity through the evolution of repair mechanisms [*refs*]. The evolvability mechanisms resulting in (quasi)Lamarckian evolution seem to have evolved jointly with and in part as a by-product of the evolution of repair" ([78] *page 5*).

The template information model does not imply that evolution toward increased complexity is perpetual, only that homeorhesis implies directionality in terms of a continuous strive toward perfect alignment with the system's 'purpose'.

However, the built-in homeorhetic properties of the template information model ensure that there is continual change. This is however, counter argued by one of the concluding statements in [9] that says "On the whole, the theoretical and empirical studies on the evolution of genomic complexity suggest that there is no trend for complexity in the history of life", and he continues with "when complexity does substantially increase, this occurs not as an adaptation but as a consequence of weak purifying selection, in itself, paradoxical as this might sound, a telltale sign of evolutionary failure." ([9] *page 1023-24*), but which from a homeorhetic perspective may be 'intentional'.

### *9.3.1 Population Genetics*

This section addresses the basic driving forces behind evolution of evolvability in terms of: a) a force adapting animals to their local environments (epigenetic regulation and plasticity) and differentiating them from each other (see sections on speciation in the previous chapter), and b) a force driving animals from simple to complex forms. The latter (i.e. the topic of bullet (b)) is described in a separate section below, so left here is the regulatory and adaptive mechanisms of epigenetics, and how this contributes to evolution of evolvability.

Romero and co-workers have a series of valuable information of relevance (see also Section 9.2.1):

- "There is also evidence that the regulation of some genes – 10-30% of genes (depending in the tissue or cell type studied)[refs] – has evolved under directional (positive) selection." ([51] *page 508*).
- Indication of species-specific as well as tissue-specific directional selection on gene regulation.
- Interspecies differences in gene expression seem only rarely to be explained by differences in alternative slicing, and only a few instances of interspecies differences in exon usage have been observed, ([51] *page 509*).
- "a substantial fraction of gene expression differences across species can be explained by inter-species changes in epigenetic mechanisms." ([51] *page 511*), for instance associated with trimethylation of histone H3 at lysine 4 (H3K4me3) – a histone mark that denotes active transcription. The authors of the study referenced estimated that up to 7% of gene expression differences across three species could be accounted for by changes in H3K4me3 status. Another study found "that as much as 12-18% (depending on the tissue) of interspecies differences in gene expression could be explained by changes in promoter methylation profiles." ([51] *page 512*). Similarly, "changes in microRNA expression levels, which are expected to affect rates of mRNA decay, could account for ~2-4% of gene expression differences across the prefrontal cortex of humans, chimpanzees and rhesus macaques[refs]." ([51] *page 512*). All in all, these types sum up to a substantial amount of the anticipated nominal gene expression.
- "… variation in methylation states between different tissues was greater than between species. Moreover, tissue-specific promoter methylation profiles were generally conserved." ([51] *page 512*).

- "…the regulation of a large subset of genes and pathways evolve under natural selection in primates [refs]." ([51] *page 508*). The references pointed at have found that "the extent of inter-species variation in gene expression levels can often be explained by variation in gene expression within a species, which is consistent with the action of stabilizing selection on gene regulation." ([51] *page 508*).

Lynch states bluntly (in terms of a heading) that "Nothing in Evolution Makes Sense Except in Light of Population Genetics" ([10] *page 8597*). He continues with "… it is reasonable to conclude that these four broad classes encompass all of the fundamental forces of evolution" ([10] *page 8597*): i) Darwin's natural selection, ii) mutations (one of which may be gain-of-function by hitherto non-functional intergenic DNA), iii) recombinations (assorting variation within and among chromosomes, and iv) genetic drift. Later in the paper, he discusses horizontal gene transfer, expressing that "The genomes of multicellular eukaryotes are invariably packed with mobile elements, and individual genes are generally subdivided by multiple introns, harbour multiple transcription-factor binding sites, and are transcribed into units containing substantial untranslated flanking sequences." and he continues "In contrast, prokaryotic genomes are usually nearly completely devoid of mobile elements and introns and have genes with very simple regulatory structures, often transcribed into polycistronic units (operons) with negligible leader and trailer sequences.", while "Most unicellular eukaryotic genomes exhibit structural features on a continuum between these two extremes." ([10] *page 8599*). This brings a new perspective to the population genetics. Horizontal gene transfer mechanisms are indeed a major player in population genetics and a major contributor to evolution of evolvability, and large enough to be described on its own – and therefore dealt with separately in the next section.

A couple of highlights from [10] are:

- "Because of their relatively small $N_g$, multicellular species are expected to accumulate gratuitous gene-structural changes without any direct selection for them and to become laden with other deleterious genomic features (refs)."[30] ([10] *page 8600*), thereby fuelling evolution with opportunities.
- "Multicellular species experience reduced population sizes, reduced recombination rates, and increased deleterious mutation rates, all of which diminish the efficiency of selection (*ref*)." ([10] *page 8600*).
- "Reductions in $N_g$ are expected to lead to increases in intron number and size, expansions in UTR lengths, losses of operons, the modularization of regulatory regions, and the preservation of duplicate genes by subfunctionalization (among other things …)." ([10] *page 8601*), and that "subfunctionalization is known to be a frequent fate of duplicate genes in multicellular species (*refs*)." ([10] *page 8601*).
- "… modular gene-regulatory structures (with unique transcription factors governing expression in different spatial/temporal contexts) can emerge passively, without any direct selection for modularity *per se*, starting from an initial state in which the entire expression beneath of the gene is under unified control (*ref*)." ([10] *page 8601*). The essence is that the gene expression re-

[30] $N_g$ is "an effective population number of gene copies", "equivalent to the effective size of a haploid population and approximately twice that for an outcrossing diploid species," ([10] page 8599).

mains constant while the underlying mechanism fix the modularity – "without a fitness bottleneck".

- "... the emergence of independently mutable subfunctions in modularized alleles can contribute to adaptive evolution in significant ways. ..." ([10] *page 8601*). One example is that the combination of subfunctionalization and their subsequent partitioning among paralogs in small to moderate populations of multicellular species provides a powerful mechanism for passive remodelling of entire developmental genetic pathways.
- "...the reductions in $N_g$ that likely accompanied both the origin of eukaryotes and the emergence of the animal and land-plant linages may have played pivotal roles in the origin of modular gene architectures on which further developmental complexity was built." ([10] *page 8601*).
- Evolution of cascades (linear pathways, in the sense that we know from the coagulation and fibrinolysis systems) may, according to Lynch, be driven entirely by non-adaptive processes of duplication, degeneration and drift, and that if augmentation of obligatory pathways are to occur in very large populations then they must be of immediate selective advantage.

In summary, population genetics has many factors to play with in the process of evolution – factors that in combination and in combination with those outlined in Section 9.2.1 and 9.3.1.1 may likely facilitate evolvability.

#### *9.3.1.1 Factors contributing to Population Genetics*

Koonin summarises a series of observations on evolutionary characteristics and on factors contributing to evolution via the population genetics:

- "The sequences and structures of genes encoding proteins and structural RNAs are generally highly conserved" ([9] *page 1015*)
- Sequences of many genes encoding core cellular functions, in particular replication, transcription, translation and central metabolic pathways, are subject to strong purifying selection ([9] *page 1016*)
- The position of a large fraction of introns are conserved ([9] *page 1016*)
- In some prokaryotic genomes, "a major fraction of genes were acquired via demonstrable horizontal gene transfer" ([9] *page 1016*) to an extent that made it "virtually ubiquitous in the prokaryotic world". Eukaryotes are different in this respect: "In multicellular eukaryotes, where germline cells are distinct from the soma, HGT appears to be rare (*ref*)" ([9] *page 1017*)
- The last common ancestor of the extant eukaryotes already possessed the mitochondrial endosymbiont (*refs*)." ([9] *page1017*)
- "eukaryotic genes that possess readily identifiable prokaryotic orthologs are sharply split into genes if likely archaeal origin (primarily, ..., components of information processing systems) and those of likely bacterial origin (mostly, metabolic enzyme and components of various cellular structures) (*refs*)" ([9] *page 1017*)
- "Indeed, in mammalian genomes, sequences derived from mobile elements, primarily, retrotransposons ... appear to constitute, at least, 40% of the genomic DNA (*ref*)." ([9] *page 1018*)

- "given that the strength of purifying selection is proportional to the effective population size, substantial increase in genomic complexity is possible only during population bottlenecks." ([9] *page 1023*)
- "Fixation of non-coding sequences, such as introns or mobile elements are, at best, neutral but, more likely, at least, slightly deleterious, ..." and therefore, "extensive accumulation of such sequences is possible only in relatively small populations in which the intensity of purifying selection falls below the 'complexification threshold'." ([9] *page 1020*)
- "the majority of the protein sequences seem to be subject to substantial purifying selection." ([9] *page 1020-21*)
- "the distribution of positively selected sites is strongly non-random among functional categories of genes, with genes involved in immunity and other defence functions, reproduction, and sensory perception being particularly amenable to positive selection" ([9] *page 1021*)
- highly expressed genes, indeed, tend to evolve substantially slower than lowly expressed genes (*refs*)." ([9] *page 1024*)
- The toxic effect of protein misfolding could suffice to explain the observed covariation of expression level and sequence evolution rate

The review of Innan and Kondrashov, [32], outlines a series of models on the evolution of gene duplications that naturally have a major impact on population genetics, divided into four categories:

I) Model assuming that the duplication does not affect fitness:

a) Ohno's neofunctionalization: when the duplicate is redundant, the "new copy can therefore be pseudogenized or lost through the accumulation of neutral loss-of-function mutations." and "as the redundant, dying gene copy accumulates substitutions, it may require a new gene function that will be maintained by selection." ([32] *page 99*); see more detail in the next section;

b) Duplication–degeneration–complementation (an extension of category I-a): also this model assumes that the new copy is redundant, but assumes the accumulation of degenerate mutation in both copies thereby reducing their efficiency; after fixation of mutations in both by drift neither copy is sufficient to perform the original function, and consequently the two copies are subfunctionalized;

c) Specialization or escape from adaptive conflict: this model assumes that gene copies are fixed by genetic drift. The essence of the assumed mechanisms is that the original gene copy was performing two functions that could not be independently improved; and if so, then the original and its copy can each be driven by positive selection to improve one of the functions and thereby specialisation happens.

II) Models assuming that the duplication itself is advantageous:

a) Beneficial increase in dosage: if an increase in expression of a particular gene is advantageous then a duplication of such gene may be fixed by positive selection. Examples of genes, where this may be advantageous are for instance: i) sensory genes, stress-response genes, and transportation genes as well as some metabolism-related genes; ii) genes with dosage-sensitive functions; and iii) genes for which their expression is needed in large quantities, such as to produce ribosomes and histone. Both the original and the copy will be under selection, but weak selection for dosage shortens the fixation phase, and "More importantly, the fate-determination phase can be longer, so that there is more chance of a fate-determining mutation occurring before pseudogenization takes place as a result of negative selection against null and deleterious mutations." ([32] *page 102*);

b) Shielding against deleterious mutations: if the two instances of the duplicated gene are at least partly redundant, then they may shield one another from the accumulation of deleterious mutations, especially in situations with increased rate of mutations. Then, if/when the one gene is lost there is still the other one as a reserve for the functionality needed;

c) Gene duplication with a modified function, where the duplication process itself creates such new function that can be fixed and preserved by positive selection; this may happen when the duplication is only partial, for instance omitting regulatory elements or other functional parts. Also, "a new genomic location of the gene copy may introduce new functional aspects[ref], or a retrotransposed copy may recruit regulatory elements in the new location or be integrated into an existing gene, which results in a chimeric gene[refs]." ([32] *page 103*).

III) Models assuming that the duplication occurs in a gene for which genetic variation exists in the population. "In some cases, these polymorphisms can immediately become fate-determining mutations that promote fixation of the duplicated copy." ([32] *page 103*):

a) Adaptive radiation model: this model "emphasizes a period of 'pre-adaptation' in the pre-duplication phase that causes adaptive fixation of the subsequent duplication." ([32] *page 104*). The example used by the originator of the model is a gene that codes for a receptor for an environmental chemical. When a similar chemical substance appears in the environment, the receptor may already have (some) ability (or will obtain it) for binding it, and in this situation gene-duplication allows for the full binding of the new chemical to be evolved without losing the old receptor function;

b) Permanent heterozygote, considering a population in which balancing selection maintains two alleles of a given gene: It is assumed that "a heterozygote for a single locus is biologically equivalent to two loci that are each homozygous for a different allele. Under these conditions, the selective benefit of the two distinct loci over one heterozygous locus is the elimination of the reduced fitness of homozygotes when only a single locus is present …" ([32] *page 104*);

c) Multi-allelic diversifying selection: In situations, where selection favours genetic variability, gene duplications are advantageous as targets for mutation and selection. "For a single-copy gene under multi-allelic diversifying selection, such as the major histocompatibility genes, overdominant selection operates. As a result, the maximum possible number of heterozygous individuals in the population is reached[ref]. Consequently, the gene that is under selection accumulates several alleles with different functions ...” ([32] *page 105*).

IV) Models in which fixation of a duplicated gene copy occurs as a by-product of other events, such as whole-genome (or chromosome) duplication and large segmental duplication:

a) Dosage balance: this model attempt to give an account for the preferential retention of some of the duplicated genes after whole-genome duplications. Consider the case of a dosage-sensitive single copy gene, A, with an optimum dosage dependency on another gene, B. There will be a negative selection against duplication of either A or B alone to avoid a dosage imbalance. Therefore, at whole genome duplication these two genes will be maintained in the population as a pair because of their interdependence, but negative selection of either alone will be deleterious because of the dosage imbalance.

Now consider the multiplicity of genes in a genome, and since the above mentioned models are not necessarily mutually exclusive, gene duplication opens a gift box of possibilities – a combinatorial explosion – for evolution to select from.

### *9.3.2 Constructive Neutral Evolution*

This theory is another controversial topic at least in the eyes of Neo-Darwinian thinking, as seen from discussions in among others: [23], [178], [179], [180] and [181]. The discussion is similar to the discussion on the Extended Evolutionary Synthesis versus the Standard Evolutionary Theory; however, the Extended Evolutionary Synthesis seems to be established now, even if there is still resistance. Similarly, time will show whether or not thee ideas in the constructive neutral evolution become acknowledged. At least, the arguments and the evidence are sufficiently convincing to include the theory here.

According to Koonin ([9] *page 1013*), “... the dominant mode of selection is not the Darwinian positive selection of adaptive mutations, but stabilizing, or purifying selection that eliminates deleterious mutations while allowing fixation of neutral mutations by drift (*ref*).”, and that “a substantial majority of the mutations that are fixed in the course of evolution are selectively neutral so that fixation occurs via random drift.” ([9] *page 1013*).

Lukeš and co-workers say that the Constructive Neutral Evolution (CNE) “is a unidirectional evolutionary ratchet leading to complexity, if complexity is equated with the number of components or steps necessary to carry out a cellular process.” ([179] *page 528*), and that it is the contingent irreversibility that serves as a neutral evolutionary ratchet, “a directional force that might drive complexification within some lineages, without positive selection.” ([179] *page 529*). This argues for its position here in the Unifying Theory of Evolution.

Stoltzfus argues that the CNE theory cannot alone explain evolution: “... thus we need not discard Darwin’s concept of natural selection as a shape-shifting force. Likewise we need not discard the concept of mutation as a mass-action force...” ([23] *page 10*), so there is no need for others to fear the new theory. The present author’s argument is that the new theory represents an innovative view on evolvability – and precisely what one should be looking for in relation to evolution of evolvability – that is, additional/’new’ mechanisms for evolving the evolution.

The 'commentary' of Doolittle and co-workers in [178] nicely summarize the essence of the constructive neutral evolution theory, while counter arguing the statements in Speijer's 'think again' paper in [180]. For instance (from [178]):

- Implementing a tendency toward complexity: "CNE is a ratchet-like process capable of generating biological complexity intrinsic to macromolecules and emphasizes the role of neutral evolution, not positive selection." ([178] *page 429*). This aspect of the CNE theory is further elaborated in [179] and [181]: gene duplication for proteins in oligomeric conglomerates enables the change of a homo-oligomer into a hetero-oligomer configuration with subsequent functional differentiation as a result of mutations, thereby achieving an "inexorable evolutionary ratchet" ([181] *page 271*). Lukeš and co-workers emphasize that "Once established, such complexity can be maintained by negative selection: the point of CNE is that complexity was not created by positive selection." ([179] *page 530*).
- "The neutral theory of molecular evolution entails that no single pre-designated neutral mutation is likely to be fixed; at the same time it holds that *some* neutral mutations inevitably will be fixed." ([178] *page 428*).
- Some changes that are either neutral (or they may even be somewhat disadvantageous to individuals) may be sufficiently advantageous to the species, for instance cultivating enhanced speciation or reduced extinction rate.
- Further, "once in place, CNE-generated complexities are preserved by negative or "purifying" selection, and may later go on to acquire useful functions." ([178] *page 429*).

Further, Stoltzfus in [23] emphasizes a couple of more aspects that may support the understanding of the theory:

- CNE is a two-step view of evolution, which by the way does not account for the entirety of evolution, also referred to as 'origin-fixation dynamics', because of the first step (duplication) and the second step (fixation by mutation and/or drift), i.e. a proposal-and-acceptance process, "whose kinetics depend directly on the introduction of novelty by mutation-and-altered-development." ([23] *page 10*).
- "… duplication creates a capacity for complementation of mutations that impair sub-functions, allowing duplicate pairs to stumble from redundancy to co-dependency, resulting in apparently specialized genes." ([23] *page 2*).
- The 'excess capacity' (the initial capacity for developmental unscrambling) is the capacity of one duplicate copy of a gene (gene product) to compensate for defects in the other. "In the CNE model for spliceosome complexification, an ancestral intron is presumed to have the gratuitous capacity to reassemble and splice when split into pieces– as shown experimentally for group II introns and even for protein-based inteins [*ref*]– , and this allows it to evolve into the multiple snRNAs of the spliceosome."[31] ([23] *page 5*).

[31] An 'intein' is "an internal peptide sequence of a protein precursor that is spliced out by transpeptidation during processing to form the mature protein." ([11]).

Doolittle and co-workers mention examples of highly significant functionality that may have been introduced by the constructive neutral evolution theory: eukaryotic sex and introns (see the authors' elaborate arguments) ([178]). Lukeš and co-workers convincingly argue in favour of the examples of RNA-editing, splicing and the ribosome ([179]). Stoltzfus further adds to this list another candidate example functionalities that may have been established by the support from the constructive neutral theory: gene scrambling besides the RNA pan-editing ([23]). In gene scrambling, the processing of gene material (in a post-mating ciliate cell) to generate the appropriately functioning macromolecular genes includes removal of intervening segments and unscrambling of the segmented genes, where the model for gene scrambling is based on a side-effect of a suggested mechanism for removing the necessary internal sequences "...using flanking "key" or "pointer" sequences, similar to repeats generated by various types of transposons." ([23] *page 5*). And as Stoltzfus state in his conclusion, "Its most provocative implication is that developmentally mediated biases in the introduction of phenotype variants represent a legitimate evolutionary mechanism ..." ([23] *page 10*).

Empirical studies show "the selection pressure on recently duplicated genes that relaxation of purifying selection was more likely to be symmetrical, to affect both duplicates more or less equally (*refs*)." ([9] *page 1022*), and from this he concludes that sub-functionalization is more likely, "whereby new paraloges retain distinct subsets of the original functions of the ancestral gene...". Later on the same page, Koonin says "organizational transitions in evolution seem to be accompanied by bursts of gene duplication, conceivably, enabled by weak purifying selection during population bottlenecks..." ([9]); such "burst" provides a potential resemblance to the mechanism of punctuated equilibria. Such bursts will explain why eukaryotes seem to have "the characteristic many-to-one co-orthologous relationship between eukaryotic genes and their prokaryotic ancestors (*ref*)." [32] ([9] *page 1022*), while giving the examples of Hox genes and other developmental regulators, and which may have had a pivotal role in the differentiation of animal phyla. Alternatively, "the burst of duplications that followed eukaryogenesis but antedates the last common ancestor of extant eukaryotes might have been brought about by the first WGD in eukarytotic evolution (*ref*)." ([9] *page 1022*, where 'WGD' is an acronym for 'Whole-genome duplication').

Missing in this description (here and earlier) – and puzzling the author – is that the molecular mechanism behind relaxing purifying selection is not clear. Until then, it seems contradictory that a selection principle can be forced to be relaxed while at the same time the selection process is still active; at least the principle inherent in (i.e. operating at the next fractal level) and beneath the natural selection must have an internal phenomenon that is not yet sufficiently explored.

### *9.3.3 Horizontal Gene Transfer and Selfish Genes*

An example of horizontal gene transfer previously mentioned is that infection with *Toxoplasma gondii* has several reported effects on behaviour, on man and more, such as effects on personality, change of fear, aggression and impulsivity; see Section 4.3.1, all of which may have an effect on evolution.

---

[32] Orthologous: common descent from a single ancestral gene. Paralogous: gene duplication relationship between genes (both definitions are from [9] (*page1015*)).

In eukaryotes, massive gene transfer from a single source is thought to be specifically associated with endosymbiosis ([182]). Qiu and co-workers express, however, that "generally, fungi … contain limited amounts of foreign genes derived from distantly related sources (e.g., [*refs]*), whereas gene transfer highways exist that allow massive gene exchanges between fungal lineages (e.g., [*refs]*)." ([182] *page 2*), and that "HGT occurs commonly between species that are in close proximity or have physical contact (e.g., [*refs]*)." ([182] *page 5*).

One of the HGT mechanisms mentioned by Koonin and Wolf in [24] – beyond those already outlined in Section 9.2.1 – is the CRISPR-Cas system, which responds directly to an environmental cue (here foreign genetic material) by introducing a genetic change into the intruding genome, and which is immediately adaptive with respect to the cue that triggered the change. The system integrates fragments of virus or plasmid DNA into a distinct, repetitive locus in the bacterial or archaeal genome ([24]). "The transcript of this unique spacer functions as a guide RNA that is incorporated into a specific complex of Cas proteins possessing DNAse activity and directs this complex to the cognate alien DNA (or RNA) molecules that are cleaved and accordingly inactivated." ([24] *page 6*). The authors compare this system with an immune system as it is extremely efficient at creating specific and targeted variation.

Many bacteria and archaea possess the transformation ability to internalize DNA from the environment ([24]). This is also scary, when one knows that DNA fragments in the soil may be incorporated in soil bacteria's genome, which may then find its way to our gut for instance through the cultivation of vegetables; think GMO.

Further, an additional apparent evolutionary mechanism involves stable phenotype modifications that are widespread in bacteria (based on epigenetic regulations) and lead to the coexistence of two distinct phenotypes in a clonal population, called 'bistability'. It is the minority group that survives in unpredictably changing environments, irrespective of the relative fitness of the two sub-populations under normal conditions ([24]). This may be an advantage for instance in scare nutritional conditions or other adverse conditions.

# 10 Integration of Phenotype within the Environment, Niche Construction

**Abstract.** Function 7, integration (see Table 2 and Table 3 in Chapter 14) is concerned with the integration of output (i.e. progeny) within the system's environment – that is, this implies interaction with the environment and which may be of a manipulative kind, for it to optimise survivability and/or generate renewed progeny – the next generation. This points at the 'niche construction theory'. The principle is 'survival of the fittest integrative capability', and the mechanism is 'natural selection of integrative skills', in this chapter explained for the three perspectives: Standard Evolutionary Theory, Inevitable Evolution Theory and Extended Evolutionary Synthesis

**Keywords:** Natural selection, survival of the fittest, Standard Evolutionary Theory, Extended Evolutionary Synthesis, Inevitable Evolution Theory, niche construction, reciprocal niche construction, technology factors

The key functionality here is the (potentially manipulative) integration of progeny within the system's environment in order to optimise survivability and/or generate renewed progeny – the next generation. However, let's take the concept a little further, to address the interaction of an individual with its environment in a wider sense.

**Principle**: Survival of the fittest integrative capability
**Mechanism**: Natural selection[33] of integrative skills
**Emergent property**: Selected phenotypes integrated within their environment, each engraved with mechanisms for engaging in future evolutionary processes

This leads to the refinement/fine-tuning of the environment essential to achieve sustainability in a long-term perspective, and to secure continued survival.

'Niche construction' is defined as "the process whereby organisms modify their own and/or each others' niches, through their metabolism, their activities, and their choices." ([34] *page 196*); see also [3]. This is also how the Mereon Matrix works (see [2]), in that the matrix expels its progeny as renewed possibilities into its environment, where the progeny is integrated in order for it to serve as new possibilities.

Further, Laland and co-workers say that "Niche construction theory replaces both proximate and ultimate causation by 'reciprocal causation' and regards the characteristics of organisms as caused by interacting cascades of selection and construction (…)." ([34] *page 200*). Obvious examples are the manufacturing of nests and burrows as hideaways or colonial residency and/or for caring of progeny, as well as webs for captivating food, and finally the caterpillars' cocoon. There are many more examples, just think of mankind's' restructuring of landscapes, innovation of technologies, domestication of other species, GMO, pollution, and much, much more.

---

[33] Note the definition of 'natural selection' in Section 1.2.

Laland and co-workers suggest that the evolutionary theory called 'developmental biology' has a lot of similarities with the 'niche construction' theory, the only difference being their application range; the one affects the system from within the system while the other affects the system via its environment ("... internal versus external environment..." ([34] *page 207*)). In a Mereon Matrix context, this distinction between invironment and environment is a highly significant difference, as Functions 1 to 6 take place within the system, while in Function 7 the progeny is immediately expelled and then integrated within its environment – that is, Function 7 takes place within the environment. Therefore, according to the Mereon Matrix these two theories take place in two different functions. Hence, their suggestion of extensive similarities is not compliant with the template information model; it is therefore strongly suggested to refer the 'niche construction' theory to Function 7, while the characteristics described for the 'developmental biology' suggest that it belongs under Function 1 in the section 'Readying for evolutionary pressure'.

"... advocates of niche construction suggest that it directs, regulates and constrains the action of selection, and is a source of evolutionary innovation..." ([34] *page 207*). Indeed it will regulate and constrain the effect of natural selection mechanisms, which it is itself prone to. Moreover, it will provide a source of evolutionary innovation, since it creates changes in the system's environment and thereby generates new conditions for future evolution. The niche construction mechanism will work on and will have an effect on the output coming from preceding natural selection steps.

Laland and co-workers in [3] express: 1) that niche construction is a process that "... directs evolution by a non-random modification of selective environments." ([3] *page 5*); 2) that the evolutionary causation is reciprocal and hence that organisms co-evolve with their environment; and further predicts that 3) "niche construction will be systematically biased towards environmental changes that are well suited to the constructor's phenotype, or that of its descendants, and enhance the constructor's, or its descendant's, fitness." ([3] *page 10*).

We strongly suggest that 'niche construction' belongs under each of the three perspectives for this function, yet with different meaning/contents.

Note here, that since this function is the concluding function on the evolutionary theories, and since it is taking place within the environment after completion of the part of the cycle that includes the evolution, one has to bear in mind the principle of sequential traversal – as seen from the spiral progression in Figure 1. This implies that the niche construction theory may exert its effects only during subsequent traversal(s) of the functions.

| **Inevitable Evolution Theory (IET)** | **Standard Evolutionary Theory (SET)** | **Extended Evolutionary Synthesis (EES)** |
|---|---|---|
| Constituent integration processes:<br>• Niche construction<br>• 'Resource enhancement altruism' | Intentional integration processes:<br>• Niche construction<br>  ○ Reciprocal niche construction | Adaptive integration processes:<br>• Niche construction<br>  ○ Reciprocity<br>  ○ Technology factors |

## *10.1 Niche Construction, 'Inevitable Evolution Theory'*

As seen from Table 1, this topic was judged relevant for this perspective here, but no dedicated literature was found specifically in relation to the Inevitable Evolution Theory and niche construction. Nevertheless, there are a couple of candidate examples.

The basic mechanism in this perspective of evolution comprises constitutive change of the environment as a result of the conditions of living and being. An example is the earthworms' consumption of dead plant material on the ground. This is obviously a transformative interaction between the earthworm and its environment, although from a food resource perspective, rather than an interaction directly dealing with survival of progeny. Nevertheless, it is a fitness aspect in that without food the worm will not survive and/or will not reproduce, and it is a niche construction aspect in that the environment changes significantly in a long term perspective and/or if there is a large population of the worms.

Similarly, a range of species, if not all, change their habitat simply by foraging and eating the food resources and returning these as excrements. It is part of the entire network of life-cycle conditions. An example is the elephant that turns over a tree to get the leaves at the top, when it cannot otherwise reach them; a simple example that in itself may not be significant in an evolutionary context apart from securing food and thereby enabling survival is a habitat with scarce resources. It is therefore obvious that organisms may unintentionally change the niches of other species within the habitat that they share, and consequently result in direct or diffuse co-evolution, and potentially with a profound effect on the stability and dynamics of ecosystems, both in macro- and micro-evolutionary timescales ([3]); just think of mankind cutting down forests to convert the area into farming area, or mankind that oust other species when populating new ground.

A further example is the larvae weaving its cocoon and thereby creating its own little safe space within its environment while transforming into a butterfly.

Local competition enhances selection for altruism, while low crowding and abundant resources eliminate the benefit of such behaviour. This van Dyken and Wade call 'resource-enhancement altruism'. Intentional and explicit modification of one's environment to increase the supply of resources is widespread in nature ([70]). Examples include provisioning behaviours, like agriculture and rearing of livestock, the latter exemplified by ants milking aphids for their honeydew.

## *10.2 Niche Construction, 'Standard Evolutionary Theory'*

The basic mechanism in this perspective of evolution comprises intentional manipulation of the environment, yet without the adaptive element resulting from the individual's learning that characterise niche construction for the perspective of the Extended Evolutionary Synthesis. It is also distinct from the niche construction within the Inevitable Evolution Theory, which has the nature of changing the environment merely as a given condition of living, for instance in the example of the rain worm that improve the soil with its excrements from digesting dead plant material.

Intentional manipulation of the environment is known from a range of microbial organisms in terms of their secretion of toxins and antibiotics into their environment as a defence mechanism.

Another example might be the birds that build a nest, and repeatedly and instinctively (i.e. without learning) build the same characteristic nest again and again even if its eggs or its small ones are taken by predators.

"Individuals from the same or different species impose selection on one another, creating a dynamically changing selective environment that evolves along with the traits that it selects." ([71] *page 2498*). The authors studied co-evolution of altruistic strategies (see these in Section 5.2.1) and found that they each have severe limitations, ultimately causing their evolution to come to a halt when their limiting factors prevails in the environment: survival and fecundity altruism increase growth rate, but are counter selected by local resource competition, while resource-enhancement and resource-efficiency altruism increase growth yield and are favoured by local resource competition. It is common in nature for a species to possess more than one of these types of altruism, especially survival/fecundity altruism and resource-based altruism; the latter are denoted 'ecosystem engineers' or 'niche constructors' because they alter the absolute resource level in their local environment, but there is no adaptive element in this kind of niche construction. It is the modification or manipulation of environmental conditions that is the criteria for calling an evolutionary trait for niche construction. Survival/fecundity altruism changes the relative resource level by increasing the local crowding and is therefore also niche construction. As the one is clearly vulnerable to scarce resources, while the other generates more resources, and oppositely the other is favoured by local crowding while the one increases local crowding, it is obvious that a synergy between the two types of altruism might be beneficial. This is precisely what van Dyken and Wade in [71] clearly demonstrate by their simulation studies: "their evolution is mutually reinforcing, creating an autocatalytic ecoevolutionary process that we call "reciprocal niche construction."" ([71] *page 2499*), this taking place indirectly via modification of the environment. They conclude by saying that "It is not a coincidence, then, that the most ecologically successful species on earth, the mound-building termites, the aculeate Hymenoptera, and humans, each possess all four altruism types identified by our theory." ([71] *page 2509*), see Section 5.2.1).

### *10.3 Niche Construction, 'Extended Evolutionary Synthesis'*

The basic mechanism in this perspective comprises adaptive processes with an emphasis on learning-based refinement of the environment – that is, acquired characteristics. A learning-based example is birds that remodel the nest (e.g. extending a narrow elongated entrance) to make it more difficult for yet another predator to get access to their progeny. Perhaps humans constitute an even better example of knowledge-based niche construction with their capability for learning from errors or insufficiencies, and for constructing high-technology artefacts.

In niche construction – for this perspective – there is a learning element in that the progeny 'inherits' its niche within the environment and are trained, actively or passively, by the parent generation, and here it takes advantage of the social learning already available in the previous micro-functions. For instance, think of the simple example of bird song that is transferred from generation to generation through learning, but which may also incorporate elements from neighbouring sources of sound such as sounds from man-made machines and songs of other bird species.

Today, hundreds of species of mammals, birds and fishes are known to learn socially; this is highly advantageous in an evolutionary context, since new acquired traits can be disseminated via the learning process and can thus increase the fitness of the individual and the group ([34]). Furthermore, these authors say that since cultural processes typically operate faster than selection processes, cultural niche construction is likely to have a more profound consequence on evolution than gene-based niche construction.

Darwin explicitly says "Man does not actually produce variability; he only unintentionally exposes organic beings to new conditions of life, and then nature acts on the organisation, and causes variability." ([20] *page 553*). This statement, a single notion within his arguments, is no longer accurate. Humans do impact the variability directly and indirectly by means of technological factors and to an extent which far exceeds breeding domesticated animals: cloning, gene manipulation, artificial and natural chemicals in agriculture (and hence in our food) and pharmaceutical drugs for disease prevention or treatment.

"All of these changes convert developmental processes from the mere unfolding of genetically guided programs to a process of active "niche regulation"." ([34] *page 201*), where niche regulation requires developing organisms to respond *partly* to inputs from their developmental environments (on the basis of their inherited genes), and *partly* to modify those same developmental environments by their niche-constructing outputs, "...based on genetically afforded (but not genetically determined) phenotypic plasticity." ([34] *page 201*). From the niche-construction perspective, a key task for any developing organism is the active regulation of its inherited 'niche', and this may be achieved in a combination of two ways: "by responding to its environment, and by altering its environment, in ways that keep its personal organism-environment relationship continuously adaptive, for the rest of its life." ([34] *page 201*).

An example of the impact of technology is clothes. The technology of fabricating clothes has enabled mankind to migrate and exploit all but the most hostile landscapes on Earth, and to evolve under these conditions.

Another example is modern western medicine. Prior to modern medicine, early onset diabetics (type I) usually died before they were able to establish a family. Today, they – as well as people with one out of a series of severe inborn genetic 'errors' like phenylketonuria – live an approximately 'normal' life with the opportunity for generating offspring, irrespective of and perhaps even in contradiction to 'fitness' in the biological sense. It is the technological advancements (in this case: medical / pharmaceutical) creating a changed environmental condition, which the individual can adapt to, rather than changes in DNA that provide their changed conditions for survival. Similarly, drugs like the antibiotics redefine the fitness for survival as they have almost eliminated the risk of death from common infectious diseases like pneumonia, and even plague or cholera. Nevertheless, today the strive for efficiency in breeding domesticated animals has led to multi-resistant strains of staphylococci, MRSA, so the technological factors once again changes our pattern of survivability, this time as a boomerang.

# Part 3: Discussion

The achievement of the present study is the unification of existing and competing evolutionary theories from the Standard Evolutionary Theory (based on a modern version of Darwin's ideas), the Extended Evolutionary Synthesis and the Inevitable Evolution Theory. From the perspective of the Mereon Matrix, they are not competitive, but complementary and cooperative theories, now placed in a framework revealing the full extent of their individual properties and how they interact.

This part corresponds to the Discussion section in traditional scientific papers and thus is concerned with the achievements in relief of the goal, as well as the strengths and weaknesses, the meaning and generalisability, as well as implications of assumptions and limitations of the study.

The hypothesis that the Mereon Matrix may be an instrument for providing the General System Theory (that von Bertalanffy called for) has definitely not been weakened by the present study, and therefore that the Mereon Matrix may indeed constitute that universal system. However, the present efforts are still considered insufficient as evidence. Formally, this can only be convincingly demonstrated when the present modelling work is near completion for a significant fraction of the 343 micro-micro-functions, accompanied by an abstraction that will release the model from its ties to the present application domain and back to a general Mereon description, and/or when other researchers have successfully repeated similar modelling studies within other knowledge domains. Also the application of the full palette of the Mereon Matrix properties may contribute to demonstrate the Mereon Matrix as a universally applicable system template and hence, an instrument for providing the General Systems Theory.

# 11 Pre-life Evolutionary Changes

**Abstract.** This chapter briefly discusses evolution of early life forms following the Big Bang.

**Keywords.** Pre-life evolution, evolution, Inevitable Evolution Theory, genesis

Gabora in [183] discusses the evolution of early life and argues for why it could not have evolved through natural selection. Even if the present study is limited to biological evolution, the topic is relevant to briefly address. The issue is how pre-life entities have evolved after the Big Bang, over the appearance of molecules to protocells that may reproduce by division, and on to a living being[34] with a system of inheritable information. Gabora suggests "that the evolution of early life is appropriately described as lineage transformation through *context-driven actualisation of potential* (CAP), with self-organized change-of-state being a special case of no contextual influence, and competitive exclusion of less fit individuals through a selection-like process possibly (but not necessarily) playing a secondary role." ([183] *page 443*).

The chapter by Woolf and Dennis in [2] provides an account for the development from the Big Bang till (early) life forms on our planet, modelled by means of the Mereon Matrix template information model, as a lineage transformation through context-driven actualisation of potential. The reader is referred the original paper for details of this process. The essence is that once the material and the right conditions were present on Earth, self-organisation would be able to drive such development. The authors' point is that self-assembly of molecules do exist (based on thermodynamic forces) and will create the basic structures that pre-life is based on until a system with genes and catalysts were able to account for the rest of the evolution.

Thus, it would be interesting to analyse whether the principle(s) and mechanisms by which such early development may have taken place are congruent with the principles and mechanisms identified here for biological evolution. A suggestion would be that it has to be evolution according to the Inevitable Evolution Theory, since initially after the Big Bang there was no genome. One such path might be the one suggested by Woolf and Dennis ([184]). Witting's Inevitable Evolution Theory may be applicable in its basic meaning to make an account for the pre-life developments taking place on our planet. Therefore, note that the general version of the principles and mechanisms (in Chapter 16) are independent of references to the genome or epigenome (the phenotype). Also note that the mechanism of F6(f4(f1) is NOT 'natural selection', but 'natural exclusion', because in this first function the evolutionary and competitive selection has not yet started, and individuals are deselected only because they are unfit for survival anyway. A conclusion on this hypothesis would require an elaboration of the model for F6(f4(f1)).

---

[34] The author abstains from attempting to define 'life' and 'living beings'.

One thing, which is a bit difficult to comprehend, is how the molecules – as they grow bigger and obtain more functionality – achieve the right conformation. Naturally, this may be explainable in terms of energetic states for the conformations in question, but might there be additional and supplementary explanations? Melkikh in [185] reviews and discusses models of the early evolution based on the presence or absence of a priori information about the evolving replicator system. He suggests a model based on the learning automata theory, which includes a priori information about the fitness space. He says "Thus, the stability of the replication process includes two different aspects: the resistance to different conformations and chemical reactions and the resistance of the pure quantum state to decoherence." ([185] *page 35*). One aspect that he briefly touches in this respect, but does not put words to, is the concept of 'chaperones'; might it be that small RNAs (that may actually self-form) might serve as such chaperones, holding the necessary a priori information even quite early within the evolution, and using its structure to secure the appropriate conformation? If so, one would be free of the anticipated insufficiency of self-organisation discussed by Gabora in [183].

Specifically for the perspective of Inevitable Evolution Theory, the forces behind the materialisation of resources, and hence also the driving forces for this perspective in F6(f4(f1)), would have to be generally applicable, because when modelling deeper and deeper all molecular synthesis would share the same physical forces within materialisation of resources. It is strictly a hypothesis that these might be the following (organised into a hypothetical next fractal level, according to the Mereon Matrix template), shaping the pre-life evolution:

- f1 Laws of physics, such as: quantum mechanics, e.g. string theory, molecular orbital theory (Schrödinger equation)
- f2 Periodic table – that is, the (coding) rules and principles behind the characteristic behaviour of individual atoms and their potential for mutual interactions
- f3 Thermodynamics
- f4 The regulatory mechanism(s) guiding the 3-dimentional structure and dynamics (the force that regulates construction in nature, of which many examples beyond crystals are mentioned in [2] (Chapter 6), such as the rhombic dodecahedral structure of honey comb cells when grown naturally, the golden ratio spiral of sea shells, the dodecahedral shape of the universe, etc.)
- f5 Forces behind differentiation, and leading to chirality and symmetry axes
- f6 Punctuated equilibrium
- f7 (not yet identified)

There is no guarantee that the solution space for the pre-life evolutionary developments – given the principles and mechanisms of the functions – are constrained to provide precisely and only those natural life forms that we know today for planet Earth; we don't even know every detail of life for planet Earth yet, for the deepest seas and caves, or for the hottest and coldest places on Earth. Further, we don't know whether life developed from scratch on Earth or was brought to Earth as microbes on an asteroid. In case of the latter, the processes behind the pre-life evolution would still be the same, yet taking place somewhere else in space.

# 12 Strengths and Weaknesses of the Work

**Abstract.** Prior to concluding on the present study, it is important to reflect on the study's strength and weaknesses, as this shall determine the degree of uncertainty to be put into the phrasing of the conclusion. This is the topic of the present chapter.

**Keywords.** Self-assessment, strength & weaknesses, bias, validity

## *12.1 Self-Assessment and Assessment of Assumptions*

The main assumption is the validity of the Mereon Matrix template information model in its published version, [2]. At this point in time such assumption still continues to be a hypothesis, since an independent frame of reference for assessment does not exist at present, and no independent research group has repeated the present study for the same or for a different application domain. Moreover, as the author has no experimental facilities at hand, it is only feasible to verify the original assumption theoretically by desk-top exploration (see Step 7 in the Methodology section in 2.2.7); this was accomplished as part of the modelling process with success.

Another assumption is that the one(s) performing the modelling is able to find the relevant information within the knowledge base of the application domain to satisfy the information need. The success in this respect is correlated with a) the ability to find the right information and/or b) the existence of actual holes in the knowledge base. There is no doubt that the knowledge base is incomplete; this is said by many, many authors, who address various scientific questions regarding evolution, and almost all authors point at future work.

So far there has been no studies that did not fit into the model or which conflicted with the model.

The set of sub-topics under the heading of 'evolution' included in the Unifying Theory is incomplete in the sense that there are indeed evolutionary concepts that are not addressed in the present study, such as 'Mendel's laws', 'Cope's rule', 'exploitative competition', 'optimal foraging theory', and more; and selection based on sex was explicitly excluded. These omissions are caused by a mere delimitation – namely caused by the concern of 'where to stop?' adding details. For the author, it was important to make plausible for every cell in the framework (i.e. evolutionary topics in Table 1) that the contents make sense and point toward future research or as topics suited for future reviews. Thus, the author intentionally stopped adding details at the point where the model was concluded to constitute a coherent wholeness without significantly sized omissions, anticipating that addition of further details would not change the structure or break the logical flow of the contents.

The Lamarckian evolution is not dealt with explicitly to the same extent as Darwin's principle for evolution. The cornerstone of the Lamarckian view was the alleged intrinsic drive of evolving organisms toward perfection, in which both variation and fixation are deterministic, [24], and leading to increased complexity. Lamarck employed two forces as drivers of evolution: 1) a force driving animals from simple to complex forms, and 2) a force adapting animals to their local environments and differentiating them from each other. The first theme above aligns with the Constructive Neutral Evolution theory. The second theme aligns with that of Extended Evolutionary Synthesis; see for instance Skinner in [36], where he concludes his review by among others with the statement "Environmental epigenetics provides a molecular mechanism for Lamarck's proposal that environment can directly alter phenotype in a heritable manner." ([36] *page 1300*). Koonin and Wolf suggest that "the Darwinian and Lamarckian modes of evolution form a continuum of evolutionary regimes defined by mechanisms of evolvability that bias mutational processes to different degrees of specificity." ([78] *page 6*). This does not conflict with the Unifying Theory of Evolution, as the Standard Evolutionary Theory and the Extended Evolutionary Synthesis are cooperating theories that both evolve incrementally based on previous output from all three perspectives.

Mendel's laws, which are concerned with the rules of inheritance of traits from each of the parents, belong outside of the Unifying Theory of Evolution, as they are concerned with inheritance through allele pairs of genes, and which happens prior to the materialisation/maturation in Function 4.

Cope's rule is already referenced by Witting in [22] and is concerned with the increase in body size over evolutionary time. There is no need to add a special section on this rule.

Sexual selection without doubt has a significant effect on evolution, and it is definitely a major topic in the literature, and excluded from the present study. Within the Unifying Theory of Evolution, it is perceived as a selection related to communicative interaction, and therefore likely would belong in Chapter 5. It is anticipated that inclusion of the topic will not change the Unifying Theory's structure or content.

'Exploitative competition' "is a form of competition in which one species either reduces or more efficiently uses a resource and therefore depletes the availability of the resource for the other species." [Wikipedia, 'exploitative competition'; accessed 27$^{th}$ Oct. 2015], which does not add new theories or major pieces of information to the Unifying Theory of Evolution; see Section 8.1.1.

In conclusion, these examples of not-included theories are not contradictory to the Unifying Theory of Evolution, and they are not changing the structure or present content of the Unifying Theory, nor are they adding significant information as judged from a simple verification. Moreover, no information on evolution was identified within the scientific literature, which conflicted with the Unifying Theory of Evolution.

### *12.2 Reflections on the Quality Criteria*

The outcome of the present study depends strongly on the quality of input resources as well as on their individual strengths and weaknesses since a synthesis of existing information is made. With reference to the quality criteria listed in the 'Methods' section for literature, there are a number of issues to be mentioned:

1) Discussions of weaknesses and strength are scarce within the domain literature, as is formal assessment of assumptions. The paper of van Veelen and coworkers ([92]) demonstrates this so clearly, see for instance their summarised

critique of the domain on their page 72. This weakens the literature as a source of information for the present study, but is perceived as a condition;

2) For the present purpose, some topics/references required an extra and dedicated effort, simply because – and in particular relevant for hot topics – the reference list was a bit old compared to the publication year so the latest news were not included, or at least the exponential growth in insight was not visible;
3) Still another topic for consideration at inclusion of articles from the literature is definition of key concepts; of course this issue is not black and white, but has a range of grey tones that also depend on 1) the ability for interpretation of the meaning; and 2) how central such concepts are applied or exploited in the paper in question and in relation to the study being performed. It was necessary to disregard a couple of papers for this reason. This shall be illustrated, but with a paper for which it was feasible to handle the uncertainty within the terminology: For instance, the paper of Ekstig ([126]) addresses the issue of 'complexity' without formally defining it apart from a casual definition: "complexity means the number of parts or the amount of differentiation among parts within an individual." ([126] *page 176*). Ekstig discusses the topic within a somewhat 360° perspective. So far so good, but a proper definition would have been a desirable outcome of the analysis; among others missing in his definition are the concepts of relationships, dynamic interactions and state spaces that would be needed according to the present definition of 'system'. The definition (also casual) by Lukeš and co-workers is a bit better in this respect: "…if complexity is equated with the number of components or steps necessary to carry out a … process." ([179] *page 528*). A better definition was not found, and complexity theory is not part of the present author's competence area, so the discussion ends here;
4) A praiseworthy impression of the literature on evolutionary theories is related to the extreme caution at stating any kind of conclusion. As said in [13], a conclusion shall reflect the level of evidence behind the individual statements; the literature on evolution theories and observations are equilibrists at this, almost always using verbs like "indicate", "suggest", "point at", "seems to", etc., and sometimes apparently irrespective of the level of evidence (in terms of substantial (number of) references) reported from.

These issues have been in focus in the present study to increase its trustworthiness. In particular, there has been emphasis on defining all key concepts throughout, in order to avoid ambiguity. The paper of Ekstig (and it is definitely not a bad paper) will be used again to illustrate how careful one has to be at defining one's key concepts when they are part of the evidence in a line of reasoning:

a) Another central concept discussed and used extensively by Ekstig in [126] in his argumentation is 'culture' – however, the concept of culture is not defined. The way he cites from the literature without a subsequent / consequential reasoning along a line of arguments and evidence makes it difficult to deduce what Ekstig in reality concludes in this respect, and the exact meaning of precisely this concept is far from obvious. Example, Ekstig cites Dennett saying "… We are the only species that has an *extra* medium of design preservation and design communication: culture. That is an overstatement; other species have rudiments of culture as well, …" ([126] *page 185*). Since Ekstig does not counter argue, we have to believe that he fully agrees with Dennett. Given the present definition of culture, a lot of advanced species actually have more than just rudiments of culture. As a reader, one is left with the feeling that there may be diverging conceptions on this concept, and therefore, the author refrain from aligning with his reasoning in the section regarding the human species as a unique animal. The disagreements are concerned with the evidence supporting some of the conclusions rather than the contents of such conclusions.

b) Another example of concepts from Ekstig in [126] that would benefit from being defined is 'extinction' of a species. He states that "... there is at present an empty space between us and other organisms, a gap previously filled by different kinds of hominids. But, being a really enigmatic fact, all these intermediate species are now extinct." ([126] *page 184-5*). Recently, it was reported that we humans all have genetic material from the Neanderthals in our genome, see e.g. [186], [187] and Section 8.2.1, meaning that there has been interbreeding and hence that we humans are not a 'pure' species, but have a share of the genomes from so-called extinct hominids. Therefore, 'being extinct' cannot mean 'the ultimate end' of that species and its various phenotypic traits, meaning that the genome is no longer available for transmission to subsequent generations as is the general understanding of that concept, but that such genome likely does not exist in a pure and complete state and cannot be retrieved. As outlined in Section 8.2.1 it is, however, not the entire Neanderthal genome that can be recovered from humans, implying that there is missing information. So, a definition of 'extinction' would have been nice to have, given such strong statement.

### *12.3 Potential Bias*

Authors' awareness of potential bias is alpha and omega when judging the accuracy of a study's conclusion, because it has to do with the level of certainty that can be assigned to the approach, to individual parts of such study, and hence to the conclusion on the outcome.

For an explorative study like the present, at least the following biases potentially are at risk (see e.g. [15]), and therefore need a discussion: local minima bias, judgemental bias, hypothesis fixation, circular inference and inclusion bias. Details of this discussion are referred to Section 18.2 in Part 4 as a whole in connection with detailed reflections on the validity of the model in general. The conclusion is as follows: These biases are well-known to the author and have been in focus throughout the study, and an analysis revealed that there seems to be no sign of serious impact from any of these biases on the outcome of the present study given its objective. Because of the constructive evaluation approach, the circular inference bias needs particular attention and seems to have been handled appropriately in the present study.

### *12.4 Reflections on the Validity of the Unifying Model*

It is not possible to prove (in the strict sense) the validity of the Mereon Matrix's template information model that was applied for the purpose of the present study; it has to be verified in a number of successive applications. Neither has it been possible to disprove the hypothesis that it was feasible to apply the template for the present modelling effort.

The analysis of the additional information on the template information model (see Chapter 17) did not contribute additional facets of the Unifying Theory of Evolution.

A couple of references additional to the ones already mentioned in all of the above discuss some of the problems, illustrating the issues at stake in many studies on evolution. For instance, the references, [170] and [176], are questioning the entire presupposition behind many models of evolution:

1) That the biological reality is far more complex than frequency-independent selection acting on scalar phenotypes (a single phenotype trait dimension) as competition and predation often leads to frequency-dependent selection, where also the current composition of a population is a factor in determining the fitness of an individual phenotype through an evolutionary feedback loop. For instance, "If birth and death rates are complicated functions of many different factors that change themselves as evolution unfolds, we do not see any reason to expect that in general, evolutionary dynamics should be simple…" ([176] *page 1371*);
2) The importance of distinguishing models for short-term frequency dynamics from evolutionary models in continuous phenotype spaces under continual input of new mutations, and over long evolutionary time scales ([176]). This is complicated by the fact that a theory or hypothesis may be demonstrated valid locally within a circumscribed application range, but is not necessarily generalizable in time or space and/or under divergent conditions;
3) The short-term simplifications often applied for the Standard Evolutionary Theory. For instance, Nishikawa and Kinjo objects the simplified models of the Standard Evolutionary Theory, concluding from their own simulation experiments that a model of cooperation between the gene-based perspective on evolution combined with plasticity can account for the cumulative selection (such as gene assimilation) in evolution, resulting in a faster evolution ([170]).

In short, any single out of the three perspectives alone is not enough to embrace the entire complexity in evolution. The present Unifying Theory of Evolution is based on a model of cooperation regarding the explanation of biological evolution. By accommodating the many facets into one unifying model, it is anticipated that the proposed model may help covering the complexity mentioned.

A couple of places, it was difficult to assign a paper to the one or the other perspective; for example, the two appearances of collective decision-making strategies. The reason is that the behind mechanism is not yet fully known. In such cases, the examples used may later move from the one to the other perspective, while being replaced by other examples that at that time will appear in the literature. At least, this will not shake the validity of the Unifying Theory as such.

Then, given that the template applied for the present application domain provided a Unifying Theory of Evolution that appears to be valid the next question is whether the same template and modelling approach is generally applicable. A consequence would be that it has to be valid as a template also in other knowledge domains. Only actual modelling attempts can verify this, but it is allowable to draw strictly hypothetical analogies. One such wild guess in physics could for instance be that Quantum Mechanics and General Relativity Theory might be two out of three perspectives, but then what is the third perspective? The present author has no suggestions, but could it be that the reason behind the inability of the professionals to explain or measure dark energy might be that there is still a physical perspective and hence a coherent set of theories that hasn't been identified? – a provocative question.

# 13 Conclusion

**Abstract.** This chapter summarises the conclusion regarding the Unifying Theory: The structure provided by the template information model enabled the unification of existing theories into one framework that embraces their qualities in a pluralistic manner - that is, a major message is that existing theories co-facilitate evolution. The Unifying Theory of Evolution consists of three independent perspectives, each incorporating seven functional levels, ranging from materialisation (readying of resources for the evolutionary pressure), communicative interaction, stabilisation (i.e. addressing efficiency), prioritization (i.e. addressing effectiveness), patterning and Time and timing (e.g. specialisation and speciation), evolvability, and integration within the environment.

**Keywords.** Unifying Theory of Evolution, evolution theories, Standard Evolutionary Theory, Extended Evolutionary Synthesis, Inevitable Evolution Theory, Mereon Matrix, template information model

The study purpose was to continue investigating the capability of the Mereon Matrix to serve as a universal system, thereby capable of modelling the domain of evolution.

In principle, it is not possible to objectively judge the accuracy of the outcome of one's own modelling efforts, because the study constitutes a constructive evaluation effort, unless applying method triangulation, but which is not feasible. There is a risk of circular inference bias if it is unconditionally concluded that the template model is valid for the present modelling purpose. However, it can be concluded that the Mereon Matrix template information model served very-well for the purpose of achieving the present study objective: It was feasible by means of the extended template information model to provide a qualitative, synthesising model that embraces existing evolutionary theories, and actually surprisingly easy.

The built-in risks of bias given by the study approach are well-known to the author and have been in focus throughout the study; and there seems to be no sign of serious impact from any of them on the outcome of the present study.

A framework has been established, in which further variants of evolutionary theories may slide in convincingly. The literature's knowledge base is not exhausted with respect to each and every theory posed, but the prevailing theories are included and a couple of random checks did not bring new or conflicting information.

It is too early at this point in time to conclude with confidence on Kirschner's hypothesis that modelling of biological systems may provide the answer to von Bertalanffy's request for a General Systems Theory (see the Introduction) and whether the Mereon Matrix's template information model would be the answer in this respect as was hypothesised early in this book.

The major message is that existing theories co-facilitate evolution. A pluralistic approach is visible in terms of the 'perspectives' of the Standard Evolutionary Theory, the Inevitable Evolution Theory and the Extended Evolutionary Synthesis, respectively.

The structure provided by the template enabled the unification of existing – and hitherto competing – theories into one framework that embraces all of their individual qualities. The Unifying Theory of Evolution consists of three independent perspectives, each incorporating seven functional levels, ranging from materialisation (readying of resources for the evolutionary pressure), communicative interaction, stabilisation (i.e. addressing efficiency), prioritization (i.e. addressing effectiveness), patterning and Time & timing (e.g. specialisation and speciation), evolvability, and integration within the environment (i.e. niche construction).

As a metaphor, think of the appearance of a person from the three perspectives, and look at them with the eyes' of a photographer: front, back and sides. The front and back perspective definitely have similarities (such as contour), but also distinguishing characteristics, like hair and face. The views of the two sides are only mirrors of each other and serve as one perspective in nearly every visual context, and very different from the other two perspectives. Still, it is one person that we are observing, but from different angles.

All of the mechanisms and observations outlined are pieces in one joint puzzle that constitute biological evolution. The benefit noticed and which resulted from the disentanglement of the three perspectives is that this structure helped immensely at an early point when the modelling felt like handling a chaotic mixture of competitive theories. Subsequently, the template model pointed at what to look for where there were missing pieces in the Unifying Theory. For some of these holes, dedicated literature surveys identified a couple of relevant articles, and for other holes in the model the content has been sketched by means of small examples from the literature at hand.

The present author perceives multiple versions of for instance population genetics, horizontal gene transfer and constructive neural evolution merely as having multiple facets tightly connected with evolution of evolvability / homeorhesis. These various aspects are separated here because of particular differences in each their founding causal mechanisms, which usually are – and with today's technology probably have to be – observed reductionistically.

In summary, the Unifying Theory of Evolution appears to be a fairly comprehensive synthesis and description of the Standard Evolutionary Theory, the Extended Evolutionary Theory, and the Inevitable Evolution Theory, embracing also all their component parts, within one shared framework.

# Part 4: Information Model

To the extent feasible, the domain model in Part 2 was separated from the documentation of the modelling process in this Part, in order to optimize comprehensibility of the individual parts for their respective target readers. This part includes a brief description of relevant parts of the Mereon Matrix and its template information model together with an elaboration of the modelling methodology. It is concerned with the modelling effort itself rather than the resulting model addressed in Parts 2 and 3.

The outcome of the modelling effort is elaboration of the template information model with one additional fractal level for each function in question; it includes: a) biological evolution, and b) general systems theory. The latter is hypothesised to lead to the General Systems Theory that von Bertalanffy called for half a century ago. Both of these additional gains formally need the accomplishment of all 343 topics to be completed, while assessment of the validity and generality likely requires completion of less than the 343 sub-functions. Requirements in this respect are discussed.

The present study is based on the template information model that is based on the Mereon Matrix; see [2]. We provided in [1] a pilot demonstration of a domain-wide model for human molecular genetics, using the Mereon Matrix as a template information model. The purpose of the present study was to continue exploring the extent to which the Mereon Matrix (in [2]) is applicable as a template for modelling the next level of detail for the extremely complex knowledge domain of human molecular genetics, in the present study delimited to a single sub-domain, biological evolution.

As one side gain, the present study will in part open the exploration of Kirschner's hypothesis mentioned in the beginning of the Introduction – saying that modelling of biological systems may provide the answer to von Bertalanffy's request for a General Systems Theory. It requires extensive modelling of one particular biological system, which is in the process of being performed as a far-sighted effort, where the present study provides only a small start. It is anticipated that it might merely require the formulation of a generalisation of the third fractal level of the template information.

There are indeed many modelling approaches to choose among, depending on the specific purpose of the modelling effort. Boogerd and co-workers discuss mechanistic modelling and the prescriptive steps that guide such efforts. Mechanistic modelling in molecular systems biology "are generally mathematical models of the action of networks of biochemical reactions, involving metabolism, signal transduction, and/or gene expression." ([12] *page 725*), either simulated numerically or analysed analytically. Such modelling efforts are numerous within the domain of evolution, where quite a large proportion of the studies on evolutionary theories identified are focussing on and exploiting numerical simulations as a means for getting insight into the evolutionary processes. One shall certainly not diminish the value of such studies, however, the application domain (molecular genetics as a whole) is extremely complex in nature while such efforts tend to address only circumscribed problem area and/or make serious assumptions – that is, a reductionistic bottom-up approach. Our pilot study had a top-down focus on a system's phenomenological behaviour, while the present study iterated between a bottom-up and a top-down approach.

# 14 The Mereon Matrix's Template Information Model

**Abstract.** Included in this chapter is an outline of relevant parts of the Mereon Matrix and its template information model together with an elaboration of the modelling methodology.

**Keywords.** Mereon Matrix, template information model, information modelling, methodology

## *14.1 About the Mereon Matrix*

Mereon Matrix is composed of two dynamic polyhedra, an outer120-faceted and an inner 144-faceted polyhedron, plus a handful of 2/3 trefoil knots. This multifaceted structure – just for the 120 polyhedron – gives rise to 3 independent axes, called A, B and C, depending on whether one looks at the three-fold, the four-fold or the five-fold symmetry; see Chapter 6 in [2]. In an information model, this would correspond to perspectives, since one is observing a single object from different symmetry axes. The structural complexity of the Mereon Matrix – which by the way is not a hard-wired physical structure – appears different along these three internal axes, but in reality it is the same object looked at from different 'angles'.

Given the template information model, there are 49 topics in our model of human molecular genetics in [1]: seven functions in a fractal model imply that the first micro level comprises the 7x7 micro-functions (i.e. those modelled in [1]). Each of these 49 micro-functions may be elaborated at the next fractal level and will then result in 343 micro-micro-functions. As the modelling of each of the 49 sub-functions is anticipated to take one to two years, or more, to complete (because one needs to acquire sufficient domain insight in every niche of the application domain) generation of the complete model will take at minimum 50 years, just to achieve the 3rd fractal level (the micro-micro-level) of the system of molecular genetics. Therefore, one has to ask whether it really is necessary to model all 343 topics of (human) molecular genetics to be convinced of the validity and generality of the Mereon Matrix template information model? This would be a Sisyphus task, because the knowledge base of the domain extends so rapidly that any subtopic would be outdated quickly after being finished. Still, each of the models of the 49 sub-domains in molecular genetics has value in themselves, like the Unifying Theory of Evolution here.

While working on all functions in an iterative and incremental fashion, a first step for the present study was to apply the qualitative universal principles of the founding model for piloting the second micro-level. In [1], Darwin's theory of evolution was dealt with in Function 6 Sub-function 4 – that is, F6(f4), and the present study therefore, elaborates the functionality of F6(f4) in the above main text, however without any reference to this specific numbering of functions and micro-functions.

The part of the Mereon Matrix that is taken advantage of in this modelling effort is described in [2] in terms of the twelve First Principles (FPs) seen in Table 2. Of these, FP4 – 11 are applied for the present modelling purpose.

**Table 2:** The First Principles of the Mereon Matrix (from [2]).

| First Principle Number | Principle |
|---|---|
| FP0 | The Mereon Matrix is perceived as a spatial-temporal nexus: a coherent pattern that is structurally independent of the external environment yet inseparable from it. |
| FP1 | The Mereon Matrix is inclusive of all forms, and diversity of form and perspectives are seen in its unity. |
| FP2 | The Mereon Matrix has a context with the capacity to set orientation, determine polarity and direct functions, and a logical core that is responsible for maintaining sustainability and ordering functionality. |
| FP3 | The Mereon Matrix requires that the local environment provides adequate resources, 'possibilities' that are simplest and stable.<br>*Keywords:* selection, recognition. |
| FP4 | **Function 1** demonstrates that contraction and specialization are essential for the acquisition, materialization, binding and initial stabilisation of resources.<br>*Keywords:* reception; pairing; foundation; containment; capacity; grounding; catalyzation; ignition; materialization. |
| FP5 | **Function 2** causes compression in the Mereon Matrix, establishing the connection critical for directed communication among all resources.<br>*Keywords:* condensation; compression; cohesion; attraction; repulsion; internal boundaries; pulse; communicative interaction. |
| FP6 | **Function 3** involves transforming resources from a minimal to a maximal state is essential to establish a full and sustainable connection within the Mereon Matrix.<br>*Keywords:* repulsion; transformation; connection; stabilization; flow; balancing; pre-utilization. |
| FP7 | **Function 4** in the Mereon Matrix facilitates ordering and redistributing resources to ensure the continuance of transformation.<br>*Keywords:* prioritization; procession, cycling; dispersion; placement. |
| FP8 | **Function 5** in the Mereon Matrix assures the coordination and orchestration of transformed resources to minimise risk and optimise quality.<br>*Keywords*: coordination; patterning; mutability; timing; multitasking; complexity. |
| FP9 | **Function 6** in the Mereon Matrix imprints resources with what is necessary to safeguard realization, sustainability and evolution.<br>*Keywords*: clarification; prediction; fine-tuning; coherence; condition; evolution. |
| FP10 | In **Function 7**, the process of generating such potential, the Mereon Matrix is invisible, obscured by 'progeny' whose 'birth' produces energetic streams of new resources that are critical to its sustainability.<br>*Keywords:* consciousness; evolutionary extension; integration; implementation; higher intelligence; endurance. |
| FP11 | Forms, functions and dynamics in the Mereon Matrix are self-referential and recursive with self-similarity occurring at the macro- and micro-levels. |

FP11 tells us that the system is fractal in nature, and it is this property that is exploited in the modelling as a whole by extending the pilot study into the present model of a Unifying Theory of Evolution.

Some of the characteristics of the applied template information model are:

1) Each function has a distinct role within the system. Its 'emergent property' comprises its end product;
2) The dynamics of the matrix implies that for individual resources a system's functions are traversed sequentially from 1 to 7. This sequential processing implies that output from one function operates as the input for the subsequent function. Thereby, the system operates as a ramp of functions, where one function takes up the challenge wherever its predecessor stopped and then adds its own contribution to the processing of the resources. At Function 7, the resources that are expelled as progeny (the emergent property) subsequently serve as renewed potential and possibilities (i.e. input) to the originating system. This way, resources within the system evolves incrementally through its cycles of processing, while the Matrix itself remains unchanged throughout time;
3) The functions are all operating all the time, thereby enabling the cooperation that is essential for the system's sustainability. Each function is distinctly different from all others, and each constitutes a necessary role within the wholeness of the system. Their operation together with their mutual interactions institute the system's organisation;
4) Since the functions are equivalent yet different counterparts, sequentially operating and cooperating there is not a leadership in the normal hierarchical sense. Similarly, there is no planning in the traditional leadership sense, since the functions' processes are perpetual and handle that which is available and in the same way each time – remember, it is the resources that are changing;
5) The fractal nature implies the question of what will be revealed when elaborating the functionality at increasingly detailed levels – that is, at lower fractal levels. In a molecular, biochemical perspective the ultimate answer will be the rules of thermodynamics in particular and the laws of physics in general; however, one should not anticipate getting anywhere near this level of detail in the present study but nevertheless one has to realise that no biochemical process will take place that does not obey the laws of physics.

All of this is illustrated in **Figure 3**. In system development terms, this diagram corresponds to a genuine waterfall model, because of the strict directionality and absence of feedback loops. In biological terms, the functionality has obvious similarities with Krebs' cycle because of cyclical nature with repeated traversals of the 7 functions that are constant and in a fixed sequence; however, the difference from Krebs' cycle is that at evolution the raw material once entering the cycle continues cycling around in the repeated cycles, and therefore, the functionality has a spiralling nature, as illustrated in **Figure 1**. In the context of biological evolution, the consequence is that the evolutionary process is infinite unless aborted by specific internal conditions (like resources becoming inappropriate for the system's functions, leading to extinction), or external conditions (like the environment becoming hostile, also leading to extinction).

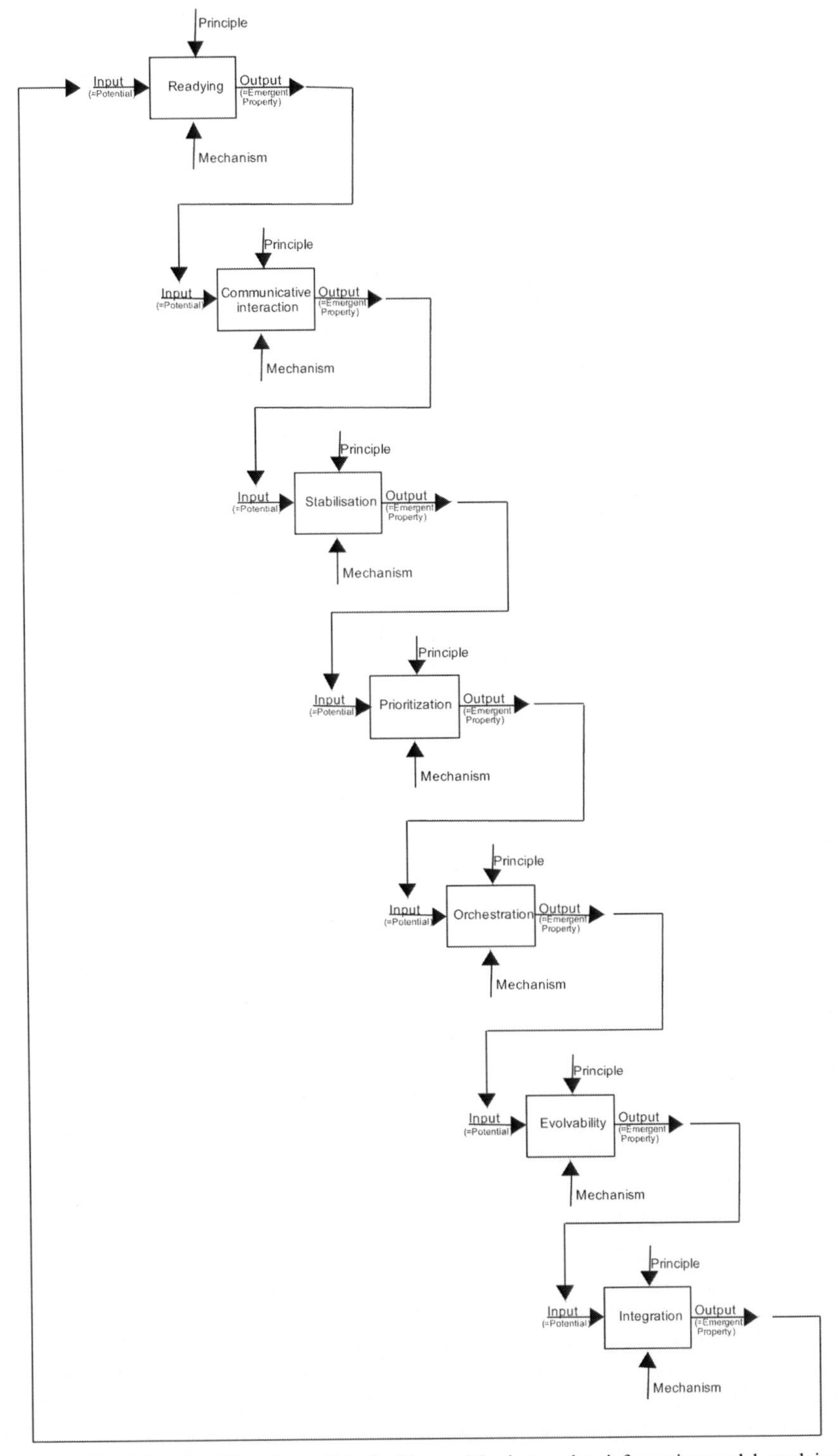


**Figure 3:** The full cycle of functions within the Mereon Matrix template information model, applying the same modified activity diagram as in Figure 2.

## 14.2 About the Modelling Methodology

This description of the modelling methodology is described in more detail than the one in Section 2.2; details that are primarily related to the modelling process rather than its outcome are added. Repetitions will appear, but only in order to make the text self-contained and thereby avoid too many referrals.

The modelling takes advantage of a template information model for modelling systems in the broad sense. The author's understanding of the concept of 'system' is based on the definition given in [2], which capture the essence of a system while making details explicit: A system is defined as "An organisation in which all structural components and dynamics are interrelational, participating internally, and affecting conditions externally" [2] (*page 480*).

Note that the modelling constitutes an incremental and iterative process – in principle the steps 2 .. 6, depending on prior insight into the knowledge domain and/or into the template information model.

### 14.2.1 Step 1: Defining the 3rd Level Mereonic Function

Since the template model includes only the macro and the 1st micro-level functionality, and since the present study aims at exploring the capability of the Mereon Matrix to serve as a universal template for information modelling, it is necessary to generate the description of the 2nd fractal level beneath the macro-level in a systematic and stringent way while iteratively and incrementally clarifying, expanding and illuminating the template model as part of the modelling process itself. That is, the study at the same time constructs the study object and evaluates it. This has similarities with constructive evaluation, and hence, the study has to be handled accordingly at the evaluation and conclusion of the study outcome, because there is a particular risk of a circular inference bias, as discussed in Section 18.2.

Based on the long generic version from Table 2 and the keywords characterising the functions, the 1st micro-level function descriptions in [1] is extended to achieve the 2nd micro level. This is then abbreviated to emphasise the essence and achieve the short version of the Mereonic function description for the 2nd micro level.

We have used styling by means of fonts and squared brackets to keep track of the different fractal levels. An example – illustrated by means of some of the results of the present modelling study is seen in the textbox:

> **F6**(f4*(f2)*)
> Mereonic function description: *Communicative interaction to facilitate* [maximising effectiveness (i.e. ordering and redistribution of resources) that **[indemnify the coherence mandatory for regeneration, sustainability and evolution]**].

The description in Italics provides the essence of the *2nd micro-level function*. The description in normal text style constitutes the 1st micro-level function. The bold text constitutes the **macro-level function**. This styling applies for both the header symbolism and the function description. The macro and 1st micro levels are taken from [1], meaning that only the information in Italics is added in this study.

For instance (from [2] *page 19*): "[F6(f4)]: Ordering the distribution of resources to safeguard the coherence mandatory for regeneration, sustainability and evolution". This was modified into: "maximising effectiveness (through ordering and redistribution of resources) **that safeguard the coherence mandatory for regeneration, sustainability and evolution**". The new version emphasizes the 'effectiveness' in Function 4 (as a consequence of the function's objectives-oriented prioritization, implying effectiveness), in order to better conceptualise the match between the Matrix formulation and the application domain's functionality.

### *14.2.2 Step 2: Getting an Overview of the Knowledge Domain*

This step is solely dedicated to activities related to the application domain, and it simply comprises multiple activities concerned with familiarising with the knowledge domain. If it were an exploration from scratch it corresponds to familiarisation with key concepts, key people, textbooks and milestone reviews on the topic, etc., depending on the application domain and the nature of the study.

### *14.2.3 Step 3: Matching the Template Model with the Application Domain's Functionality*

We take a similar approach as for Step 1 in determining the link between template functions and the corresponding domain functions: interpreting the topic of the subfunctions from [1], using not only the function descriptions, but also continue taking advantage of the keywords of the Mereonic functions as important instruments.

Application of the template information model is always a process of interpretation while matching the known functionality and characteristics of the Mereon Matrix with those of the knowledge domain. This step constitutes a translational activity, where the keywords and the interpretations of the application domain's functionality shall lead to a fairly precise layout of the content, although maybe not yet to the level where a formulation of 'principles' and 'mechanisms' are clear. If not, these have to be formulated as part of Step 4 or in a next iteration.

Depending on the modelling task, various aspects of the Mereon Matrix may be relevant to involve. In [1], we mainly applied the function descriptions and the fractal property. The three perspectives used in this study may be extremely valuable if one has sufficient detail for the system modelled, and probably in particular valuable if the wealth of information includes seemingly conflicting information as for the present study. Here, the initial interpretation 'data-driven', 'information-driven' or knowledge-driven' helped distinguishing the perspectives and which specific perspective the topic of a given literature paper (i.e. a theory or a piece of information) belongs to. However, it was also necessary to transcend that meaning of the categorisation, because the perspectives have much deeper meaning than that, as seen from later additions to the keywords categorising the perspectives.

The approach for identifying the placement of the individual (sub)-theories/topics within the Unifying Theory was to look carefully at the words used by the authors of a paper in order to characterise the process/mechanism taking place; in this respect the extensive use of citations helped securing the original meaning applied by the authors themselves.

The above example is continued in the below textbox – while again applying the approach of styling:

In an evolutionary context, this micro-function is concerned with the *mechanisms (inherently residing within the resources themselves) for communicative interaction among agents to facilitate* [mechanisms that maximise effectiveness and thereby increase **[the capability of evolutionary potentials that enable future phylogenetic differentiation]**].

In short, the focus of this sub-function is:

...*directed communicative interaction (f2)*
...for the ordering and redistribution of resources (prioritization) (f4)
...**that indemnifies phylogenetic differentiation** (F6)

The context here for interpreting the task of *f2* is its contextual macro-function, – that is, communicative interactions (*f2*) in relation to the endeavour toward effectiveness (f4) of evolution (**F6**).

'Needs and preferences' refer directly to **F6**(f4)'s focus on effectiveness – that is, doing the (perceived) right things. The communicative interaction comes into action in terms of the relationship and interactions among a set of agents. Thus, this micro-function is concerned with the (evolutionary optimisation of) relationships between agents.

**Principle**: Survival of the fittest individual

... where the 'communicative interaction' engages social and behavioural needs and preferences, i.e. kin-to-kin interactions *(f2)*

... where aspects of 'regulation' implicates the prioritization that secure 'effectiveness' (f4)

... together they provide parts of the design for maintaining the 'sustainability, regeneration and evolution' necessary for management of the change that constitutes evolution (**F6**)

**Mechanism**: Communicative interactions guiding natural selection among kin

**Emergent property**: Selected phenotypes

The '**Principle**' corresponds to the essence of the function: "Principle ... **7**. a rule or law concerning a natural phenomenon or the behaviour of a system..." ([5]).

The '**Mechanism**' corresponds to the implementation approach for achieving the Principle: Boogerd and co-workers express the purpose of spelling out a mechanism as "... how some phenomena of interest–some reliably generated behaviour of the system–is generated by reference to how a number of components interact." ([12] *page 727*). The authors bring two definitions from the literature: "Mechanisms are entities and activities organized such that they are productive of regular changes from start or set-up to finish or termination conditions" and "...a structure performing a function in virtue of its component parts, component operations, and their organization. The orchestrated functioning of the mechanism is responsible for one or more phenomena ..." ([12] *page 727*).

Note that the 'Principles' and 'Mechanisms' together constitute elements of the General Systems Theory for the present part of the entirety (i.e. functions **F6**(f4*(f1)* ..

**F6**(f4*(f7)*)) but only when they have been transformed from the application domain's terminology into general terms.

Also note that several of the comments in the above textbox are working notes dedicated to support the interpretation of the functionality, and specifically kept in order to demonstrate the modelling process.

### *14.2.4 Step 4: Filling Details for the Micro-Micro-Level Functions*

The present step is dedicated to activities related to the application domain: filling in details of the application domain's knowledge, formulated within the domain terminology.

The seven micro-functions of a given function – that itself may be a micro-function of a higher level function – cover the exact same topic(s) as the function but at an elaborate and dedicated degree of detail.

### *14.2.5 Step 5: Defining the Emergent Properties*

The '**Emergent Property**' outlines the outcome of the function's operation. Actually, the definition in [2] (*page 540*) ("Arising at each stage of a system's function a capability or characteristic change occurs, imprinted in the resources") didn't help as a definition within the present context. Instead the dictionary, [5], provided a valuable understanding: "property … **6**. a quality, attribute, or distinctive feature of anything…" with the adjective 'emergent' adding the meaning "coming into being or notice".

The Principle, Mechanisms and Emergent Property for each sub-function constitute the wholeness describing a function. The important regarding the emergent property is that every (micro-) function elaborates on that which enables subsequent functionality. One example is that histone modifications enables coding for the alternative splicing (histone modifications constitute an F1 functionality, materialisation; coding constitutes an F2 functionality, communicative interaction; and alternative splicing constitutes an F4 functionality, regulation/prioritization).

The emergent property for a given 1$^{st}$ micro-function (e.g. that of **Fx**(f1)) is copied without change to its 2$^{nd}$ micro-level function – that is, to **Fx**(f1*(f7)*)). Following this, the emergent properties of **Fx**(f1*(f1..f6)*) are open for being determined in accordance with the Principles and Mechanisms and the progression along the sub-functions.

### *14.2.6 Step 6: Identifying Holes in the Model*

Are there missing or insubstantial pieces of information within the model? This may be caused by for instance incomplete domain knowledge, an incomplete literature search, and/or a flaw in the template information model. The purpose of this step is 1) to identify the source of potential holes observed and – if feasible – remedy such flaws either by an additional information search or in the template model, and 2) to analyse whether additional information in the template model will help providing more details to the model in order to eliminate potential flaws at an earlier point in the modelling process.

### *14.2.7 Step 7: Evaluation of the Model*

Modelling based on the Mereon Matrix is judged to be successfully accomplished in case the following are fulfilled: a) when there is an adequate match at macro and micro levels between data and information of the knowledge domain and the properties of the

Mereon Matrix; and b) when functions 1..7 are sequentially addressed at both macro and micro levels and at a sufficient level of detail. Consequently:

1) Verify that the functionalities included are consistent with the functions of the Mereon Matrix and each their keywords;
2) Walkthrough all of the sub-functions in the sequence **Fx**(fy)→**Fx**(fy*(f1)*) .. **Fx**(fy*(f7)*) → **Fx**(fy+1*(f1)*), and so on - for all **x** and y that have been elaborated, in order to verify appropriate coherence, continuity and completeness;
3) Like for cross-words or Sudoku – every 'row' and every 'column' of the model has to 'add up' and contribute appropriately for the wholeness to be valid. The approach in the present study was to continuously make working notes in terms of for instance tables of the principles, mechanisms and emergent properties to ease the overview, support the modelling, and assess the coherence and completeness;
4) Verify that each function's emergent property serves as the appropriate resource input required by its successor function at both the macro and micro-level;

   Comment: Another approach for this kind of verification could be to apply diagramming techniques that shows input, output and determinant agents with feedback loops for where they are used in a next traversal of the function;
5) Is the modelling of a given micro-micro-function accomplished and concluded appropriately? Are the concepts elaborated in terms of real functionality? What is the coverage compared with the functionality for the 1st micro-level function; so carry out a walk-through to see what is not addressed – to see whether there is a need to elaborate the 2nd micro-level, or maybe something is misplaced and hence needs to be moved to another place in the model?
6) Are there any chunks of information in the knowledge domain that has not found its place in the model? Or oppositely, are there significant holes in the model?
7) There should be a minimum – if any – of chicken & egg problems. We do have a chicken & egg problem a couple of places in the entire model in [1], however, justifiable as discussed in the reference, simply because the modelling is looking at the system for a given timeslot. One has to be aware of this phenomenon and assess when and why a chicken & egg problem is justified and when not;

   Comment: For example, in [1], we use enzymes in functions **F1**, **F2** and **F3** before enzymes are defined in **F3**(f3*(f3)*) – here one has to remember that we were and still in the present study are looking at a time slot rather than the evolutionary perspective. This means that when observing an individual being at a particular point in time, there will be generations after generations of previous genetic and epigenetic changes. At any point in time that is particularly observed (here: the course of evolution), evolution has access to the entire genotypic and phenotypic functionality (including the epigenotype) developed so far, and which are available, including their accumulated changes that serve as new potentials in subsequent traversals of the seven functions.

8) Assessment of the validity of the resulting model (concepts adopted and adapted from [188]; for more, and more detail, see [189] – or even Wikipedia): Aspects of validity include for instance:
    - Construct validity (does one really measure that which one believes or intends)
    - Internal validity (degree of compliance between the perceived meaning and the reality, i.e. with minimal bias)
    - External validity (generalizability to other contexts of investigation)
    - Empirical validity (accuracy toward the true value of a measurement)
    - Rational validity (coverage or representativeness of characteristics)
    - Reliability (consistent outcome)

    Comment: When one becomes aware of a problem in this respect, the solution is to phrase the strength of the conclusion accordingly, for instance making a potential risk explicit and/or phrasing the (un-)certainty regarding the conclusion with caution (i.e. with weaker words).

### *14.2.8 Step 8: Assessing Supplementary Information from the Mereon Matrix*

Given that the book on the Mereon Matrix (i.e. [2]) comprises some 280 pages of condensed information on the matrix's structure and function in mathematical and geometrical terms, there will inevitably be supplementary information on the Mereon Matrix that is not addressed in the present modelling study. So, one should not expect that it will be feasible to make correlations between all information in the mentioned book and the different aspects in the Unifying Theory of Evolution. Examples of supplementary information are the remaining First Principles, as well as the genesis, jitterbugging, quantitative aspects, breathing and birthing aspects, and last but not least the structural aspects. These are briefly addressed in Chapter 17.

# 15 The Overall Mereon Matrix Model of Human Molecular Genetics

**Abstract.** This chapter provides an overview of the model of human molecular genetics in [1] with the extension regarding biological evolution (and more), thereby providing a larger context for the present study.

**Keywords.** Human molecular genetics, Mereon Matrix, evolution, template information model

An overview of the model of human molecular genetics (in [1]) with the extension regarding biological evolution (and more) is provided in Table 3, thereby this chapter – from a modelling perspective – provides a larger context for the present study.

So far included are the descriptions, yet not elaborated at the 2nd micro-level, for processes like replication, transcription and translation, cell signalling, cell division and cellular differentiation, as well as the foundation for evolution within the cellular dynamics at the macro and first micro-level of the functions. At present, topics like theories of evolution, enzyme kinetics, and principles behind homeostasis and homeorhesis, and more, seem to naturally slide in at the 2nd micro-level.

**Table 3:** Illustrative examples of the macro and micro-functions for a model of human molecular genetics, accomplished by means of the Mereon Matrix template information model. The first two levels (leftmost two columns) are taken from [1], while the third level items comprise highlighted pieces of information from the ongoing investigation. The bold sub-functions are the ones dealt with in this study.

| Macro-level function | 1st level micro-function | 2nd level micro-function | Name of functionality |
|---|---|---|---|
| F1 | | | Materialisation of the genome |
| | F1(f1) | | Materialisation of the genome sequence |
| | F1(f2) | | Information carrying ability of the genome |
| | F1(f3) | | Functional relations between conformation and efficiency |
| | F1(f4) | | Means for regulating accessibility of the genome |
| | F1(f5) | | Fidelity of the genome materialisation |
| | F1(f6) | | Diversity of the genome |
| | F1(f7) | | Integration of the genome into the reproductive nuclear location |
| F2 | | | Communicative transactions |
| | F2(f1) | | Materialisation of communicative agents |

| Macro-level function | 1st level micro-function | 2nd level micro-function | Name of functionality |
|---|---|---|---|
| | F2(f2) | | Coding of genomic information |
| | F2(f3) | | Efficiency of the genome's communicative interaction |
| | F2(f4) | | Effectiveness of the genome's communicative interaction |
| | F2(f5) | | Fidelity of the communicative interaction |
| | F2(f6) | | Diversity in mechanisms for the communicative interaction |
| | F2(f7) | | Implementation and integration in relation to the communicative interaction |
| F3 | | | Gene Expression |
| | F3(f1) | | Materialisation of gene expression |
| | F3(f2) | | Communicative interactions in gene expression |
| | F3(f3) | | Efficiency of gene expression |
| | | F3(f3(f3)) | Enzyme kinetics |
| | F3(f4) | | Effectiveness of gene expression |
| | F3(f5) | | Fidelity of the genome expression |
| | F3(f6) | | Diversity in mechanisms for gene expression |
| | F3(f7) | | Implementation and integration of gene expression |
| F4 | | | Prioritized gene expression |
| | F4(f1) | | Materialisation of prioritized gene expression |
| | F4(f2) | | Communicative interactions in prioritizing gene expression |
| | F4(f3) | | Efficiency of the prioritized gene expression |
| | F4(f4) | | Effectiveness of the prioritized gene expression |
| | | F4(f4(f3)) | Regulating for efficiency |
| | | F4(f4(f4)) | Regulating for effectiveness through coupling of dynamically regulated networks |
| | F4(f5) | | Fidelity of the prioritization of gene expression |
| | F4(f6) | | Diversity in mechanisms for prioritization of gene expression |
| | F4(f7) | | Implementation & integration of prioritized gene expression |
| F5 | | | Timing as process at cellular creation, re-creation, and differentiation |
| | F5(f1) | | Materialisation of cellular creation, re-creation, and differentiation |
| | F5(f2) | | Communicative interactions at cellular creation, re-creation, and differentiation |
| | F5(f3) | | Efficiency of cellular creation, re-creation, and differentiation |

| Macro-level function | 1st level micro-function | 2nd level micro-function | Name of functionality |
|---|---|---|---|
| | F5(f4) | | Effectiveness of cellular creation, re-creation, and differentiation |
| | F5(f5) | | Fidelity of cellular creation, re-creation, and differentiation |
| | F5(f6) | | Diversity in mechanisms for cellular creation, re-creation, and differentiation |
| | F5(f7) | | Implementation and integration of cellular creation, re-creation, and differentiation |
| F6 | | | Phylogenetic differentiation – diversity as a means for evolution |
| | F6(f1) | | Materialisation of phylogenetic differentiation & diversity |
| | F6(f2) | | Communicative interactions at regeneration & evolution |
| | F6(f3) | | Efficiency of re-generation and evolution |
| | F6(f4) | | Effectiveness of re-generation and evolution |
| | | F6(f4(f1)) | **Materialisation: Readying for the Evolutionary Pressure, Survival of the Fitted** |
| | | F6(f4(f2)) | **Communicative Interaction: Contributing Relational Aspects to Evolutionary Processes** |
| | | F6(f4(f3)) | **Stabilisation: Balancing Efficiency and Effectiveness at Evolution** |
| | | F6(f4(f4)) | **Prioritization: Maximising Effectiveness of Evolution** |
| | | F6(f4(f5)) | **Differentiation: Survival of the Fittest Orchestration** |
| | | F6(f4(f6)) | **Evolution of Evolvability, Survival of the Fittest Evolutionary Mechanisms** |
| | | F6(f4(f7)) | **Integration of Phenotype within the Environment, Niche Construction** |
| | F6(f5) | | Fidelity of Regeneration and Evolution |
| | F6(f6) | | Learning as a means to achieve regeneration, sustainability and evolution |
| | F6(f7) | | Implementation and integration of mechanisms for evolution |
| F7 | | | Presentation (and integration) of phenotype within the global environment |
| | F7(f1) | | Materialisation of progeny and product |
| | F7(f2) | | Communicative interactions at release & integration of progeny and product |
| | F7(f3) | | Efficiency of progeny's release & integration |
| | F7(f4) | | Effectiveness of progeny's release & integration |
| | F7(f5) | | Fidelity of release & integration of progeny and product |

| Macro-level function | 1st level micro-function | 2nd level micro-function | Name of functionality |
|---|---|---|---|
| | F7(f6) | | Diversity in mechanisms for release & integration of product & progeny |
| | F7(f7) | | Realisation of the release & integration of progeny & products into the global environment |

# 16 Extending Parts of the Template Information Model

**Abstract.** The template information model in [2] includes the function description (corresponding to the macro-level) and the initial elaboration of the 1st fractal level (corresponding to the first micro-level). The purpose here is to develop the 2nd micro-level of the template information model, however delimited to include only the function addressed in this study. This is accomplished by systematic elaboration of the initial micro-level description for the model on human molecular genetics to match the characteristics identified in the knowledge domain, followed by reverse transformation of results into a generic system version.

**Keywords.** Template information model, fractals, Unifying Theory of Evolution, information model, principle, mechanism, emergent property

The starting point for the present modelling effort is the following information from [1] (*page 487*):

> "**F6**(f4)**:** Effectiveness of Re-Generation and Evolution
> From a modelling perspective this constitutes the micro-level Function 4 (prioritization) of Function 6 (evolution)"

Function 6 and Function 4 both designate change as built-in traits, which immediately suggest that this micro-function is related to evolutionary aspects. 'Prioritization of evolution' provides a hint toward a principle for selection of survivors, namely through "Ordering the distribution of resources to safeguard the coherence mandatory for regeneration, sustainability and evolution" ([1] *page 487*). The description continues with:

> "**F6**(f4) … is made up of those mechanisms that maximise effectiveness and thereby increase the capability of evolutionary potentials that enable future phylogenetic differentiation.
>
> The **emergent property** is resources engraved with mechanisms for generating diversity as a foundation for establishing evolution. This leads to the refinement/fine-tuning of the resources essential to achieve sustainability in a long-term perspective, and to secure continued survival through coherent genomic evolution.
>
> In this micro-function, the focus is on those specific mechanisms behind the effectiveness of evolution, i.e. regulating the capability (~competence) of evolutionary potentials; is it the right kind of potentials? Important here is the capability for regeneration, sustainability and long-term viability."
>
> (end of citation from [1])

From the definition of 'system' in Section 14.2, the elements within the system are seen: structural components, dynamics, interrelations, and interactions with the external environment.

The 'structural components' come into play during the materialisation/readying of resources – that is, according to the template model this takes place in **F6**(f4*(f1)*). The actual genomic changes take place elsewhere within the system of molecular genetics.

At the output end of function **F6**(f4) – that is, from the definition of 'system' it is equally obvious that the 'interactions that affect conditions externally' are at play in terms of the integration of output within the environment, i.e. **F6**(f4*(f7)*), as shall be seen later in this chapter.

Further, from the definition of the Mereon Matrix functions in Table 2, we see that the dynamics at **F6**(f4*(f2..f4)*) is concerned with the interrelations between involved agents (i.e. resources in the template terminology), while addressing respectively the communicative interactions, the pre-utilization of resources, and the system's internal regulation / prioritization.

Left are then patterning/fidelity/time & timing (**F6**(f4*(f5)*)) and evolution of the topic at hand (**F6**(f4*(f6)*)).

Such matching of key aspects in the knowledge domain with key aspects of the functions in the Mereon Matrix generates a quick overview of the micro-level functions that together constitute function **F6**(f4).

A general point before starting is that this description is the concerned with how the original function descriptions in [2] are interpreted to something meaningful specifically in the context of the object modelled: For instance, the phrasing for Function 4 is "**Function 4** in the Mereon Matrix facilitates ordering and redistributing resources to ensure the continuance of transformation" (see Table 2); the 'transformation' addressed is concerned with the change in the resources. The original and universally applicable formulation from [2] ("…ordering and redistribution of resources (prioritization) …" is shortened to "maximise effectiveness (prioritization)", thereby emphasizing the goal-orientation of this function's operation. Such re-writing of the original function descriptions is necessary in order to facilitate the identification of the link between the template and the application domain's concepts, and naturally, such re-writing must not significantly change the meaning of the template description.

Furthermore, to understand the template information model, it is important to realize that there are various ways to express the functionality, simply because our language is too inefficient to be able to capture the entirety of a function in terms of only a single short statement. Therefore, the variation in formulation may facilitate an understanding of nuances in context. Here, the keywords are particularly helpful.

### *16.1* **F6**(f4*(f1)) Readying for the Evolutionary Pressure: Survival of the Fitted*

From Table 2:

> **Function 1** demonstrates that contraction and specialization are essential for the acquisition, materialization, binding and initial stabilisation of resources.
>
> *Keywords*: reception; pairing; foundation; containment; capacity; grounding; catalyzation; ignition; materialization.

It is the "materialisation, binding and initial stabilisation of resources" from the template description that is the focus in the present context, as the resources are not acquired from scratch.

Mereon Matrix's template description for ***F6***(f4*(f1)*) (compressed):

*Resources are readied to facilitate* [maximising effectiveness (i.e. ordering and redistribution of resources) that **[safeguard the coherence mandatory for regeneration, sustainability and evolution]**].

When translated to the evolutionary context, this micro-function is concerned with *the establishment / readying of the resources' properties in terms of a state or conditions to facilitate* [mechanisms that maximise effectiveness and thereby increase **[the capability of evolutionary potentials that enable future phylogenetic differentiation]**].

In short, in the context of evolution, the focus of this micro-function is:

*...readying of resources (f1)*
...to maximise effectiveness (prioritization) (f4)
**...that indemnifies phylogenetic differentiation (F6)**

All the mutations, relocations, etc., serve as parts of the foundation for evolution and arise outside the scope of the present study. Some arise during mitosis or meiosis, and others again at regulatory processes that fail. The key point of the present micro-function is the readying of the individual organism so that it may partake in the evolutionary processes:

**Principle**: Survival of the fitted
... where the readying of resources includes maturation *(f1)*
... where aspects of 'regulation' implicates the prioritization that secure 'effectiveness' (f4)
... together they provide parts of the design for maintaining the 'sustainability, regeneration and evolution' necessary for management of the change that constitutes evolution **(F6)**
**Mechanism**: Natural exclusion of unfit phenotypes during maturation of phenotype traits
**Emergent property**: Selected, matured phenotypes

The likelihood that external factors will impose an evolutionary process of selection on the resources is definitely present; see for instance the concept of transplacental epigenetics in [38]. Another example is the culture-based intentional abortion of female foetuses that is common in some cultures/countries.

For general systems, **F6**(f4*(f1)*)'s Principle, Mechanism and Emergent Property are:
***Principle:*** Survival of resources fit for partaking in evolutionary processes
***Mechanism****:* Natural exclusion of unfit resources
***Emergent property****:* Selected matured resources in a homeostatic state

### *16.2* **F6**(f4*(f2)) Communicative Interaction Contributing Relational Aspects to Evolutionary Processes*

From Table 2:

**Function 2** causes compression in the Mereon Matrix, establishing the connection critical for directed communication among all resources.

*Keywords:* condensation; compression; cohesion; attraction; repulsion; internal boundaries; pulse; communicative interaction.

The focus of this function is on "connection critical for directed communication among all resources" – that is, purposeful and meaningful interactions among resource.

Mereon Matrix's template description for ***F6***(f4*(f2)*) (compressed):

*Communicative interaction to facilitate* [maximising effectiveness (i.e. ordering and redistribution of resources) that **[safeguard the coherence mandatory for regeneration, sustainability and evolution]**].

In an evolutionary context, this micro-function is concerned with *the mechanisms (inherently residing within the resources themselves) for communicative interaction among agents to facilitate* [mechanisms that maximise effectiveness and thereby increase **[the capability of evolutionary potentials that enable future phylogenetic differentiation]**].

In short, in the context of evolution, the focus of this micro-function is:

*...directed communicative interaction (f2)*
...to maximise effectiveness (prioritization) (f4)
**...that indemnifies phylogenetic differentiation (F6)**

The context here for interpreting the task of *f2* is its contextual macro-function – that is, communicative interactions *(f2)* in relation to the endeavour toward effectiveness (f4) of evolution (**F6**). The communicative interaction comes into action in terms of the relationship and interactions among a set of system components. Thus, this micro-function is concerned with the (evolutionary optimisation of) relationships between individual agents.

**Principle**: Survival of the fittest individual

... where the 'communicative interaction' engages social and behavioural needs and preferences, i.e. kin-to-kin interactions *(f2)*
... where aspects of 'regulation' implicates the prioritization that secure 'effectiveness' (f4)
... together they provide parts of the design for maintaining the 'sustainability, regeneration and evolution' necessary for management of the change that constitutes evolution **(F6)**

**Mechanism**: Communicative interactions guiding natural selection among kin
**Emergent property**: Selected phenotypes

"Needs and preferences" refer directly to **F6**(f4)'s focus on effectiveness – that is, doing the (perceived) right things.

The role of selection is to promote certain advantageous parts of the gene pool, be it alleles, phenotype traits or a full phenotype, through selection based on the ability to survive and/or generate surviving progeny that then carries the advantageous genes. Here, in F6(f4(f2)), the selection focus is on the interaction between individual phenotypes, favouring not only the most fit individual, but also the one that interacts appropriately with kin.

For general systems, **F6**(f4*(f2)*)'s Principle, Mechanism and Emergent Property are:

***Principle**:* Survival of the fittest individual resources

***Mechanism**:* Communicative interactions guiding natural selection at the level of individual resources

**Emergent property**: Selected resources

## *16.3* **F6**(f4*(f3)) Balancing Efficiency and Effectiveness of Evolution*

From Table 2:

> **Function 3** involves transforming resources from a minimal to a maximal state is essential to establish a full and sustainable connection within the Mereon Matrix.
>
> *Keywords:* repulsion; transformation; connection; stabilization; flow; balancing; pre-utilization.

The "maximal state" together with the keywords 'pre-utilisation' and stabilisation' signifies a strive toward efficiency and homeostasis.

Mereon Matrix's template description for ***F6***(f4*(f3)*) (compressed):

*Maximising efficiency and stability of resources to facilitate* [ordering and redistribution of resources that **[safeguard the coherence mandatory for regeneration, sustainability and evolution]**].

In an evolutionary context, this micro-function is concerned with *regulating the efficiency while creating and maintaining a state of homeostasis to facilitate* [mechanisms that maximise effectiveness and thereby increase **[the capability of evolutionary potentials that enable future phylogenetic differentiation]**].

The phrasing "ordering and redistribution of resources (prioritization) ..." is the original and universal formulation from [2]. This is shortened to "maximise effectiveness (prioritization)", thereby emphasizing the goal-orientation of this function's mode of operation.

In short, in the context of evolution, the focus of this micro-function is:

*...maximising efficiency and stability of resources (f3)*

...to maximise effectiveness (prioritization) (f4)

**...that indemnifies phylogenetic differentiation (F6)**

The combination of "maximising efficiency" (i.e. 'doing the things right') in relation to "effectiveness" (i.e. 'doing the right things') is concerned with enhancing the capacity and capability for 'doing the right things right' while founding stability. Optimisation always comes at a cost somewhere somehow internally within the system. The definition of 'system' explicitly tells us that a system has an internal structure; that is, the system has a wholeness, and it has component parts. Therefore, it is obvious that the necessary regulation to achieve optimisation may have the consequence of prioritising that which is beneficial either for the system as a whole or for specific parts of its internal structure at the expense of other parts. In the context of evolution, this points at the situation of individual resources within a group structure of related resources.

**Principle**: Survival of the fittest group of individual resources

... where the 'maximisation of efficiency and stability' entails regulation of capabilities *(f3)*

... where aspects of 'regulation' implicates the prioritization that enable 'effectiveness' (f4)

... together they provide parts of the design for maintaining the 'sustainability, regeneration and evolution' necessary for management of the change that constitutes evolution **(F6)**

**Mechanism**: Natural multi-level selection
**Emergent Property**: Selected phenotypes

A group comprises interacting individuals, and those individual resources are also individually exposed to the evolutionary selection described for **F6**(f4*(f1)*) and **F6**(f4*(f2)*), so the evolutionary pressure in **F6**(f4*(f3)*) for a group with its individuals comes on top and is intertwined with the individual selection – therefore, the term 'multilevel selection'.

The First Principle 11 of the Mereon Matrix (see Table 2) shows us that the system is fractal, which one may observe as systems within systems. Groups in the sense that the literature on evolutionary theories uses this concept are similar to such sub-systems in that groups are self-contained entities with internal structure. When thinking of groups in the context of evolution of biological beings, such groups have dynamics, their components parts are interrelational, acting internally, and also affecting conditions externally (when that is relevant from a function perspective). The difference between a system and a group is that 1) a 'group' may dynamically change in various contexts – which a system does not; 2) a group may have overlapping sub-groups, while a system cannot have overlapping sub-systems; and 3) there is no requirement for a group with respect to discrete functions or roles, which is the case for systems.

For general systems, **F6**(f4*(f3)*)'s Principle, Mechanism and Emergent Property are:
***Principle:*** Survival of the fittest group of individual resources
***Mechanism:*** Natural multilevel selection
***Emergent property:*** Selected resources

### *16.4* **F6**(f4*(f4)) Maximising Prioritization for Effectiveness of Evolution*

From Table 2:

> **Function 4** in the Mereon Matrix facilitates ordering and redistributing resources to ensure the continuance of transformation.
>
> *Keywords:* prioritization; procession, cycling; dispersion; placement.

The focus is on "ordering and redistribution to ensure the continuance" (i.e. a strive toward 'effectiveness') together with the keywords 'prioritization' and 'cycling' that tells a bit about the means or process to fulfil the strive toward effectiveness.

Mereon Matrix's template description for ***F6***(f4*(f4)*) (compressed):

*Maximising long-term effectiveness to facilitate* [ordering and redistribution of resources that **[safeguard the coherence mandatory for regeneration, sustainability and evolution]**].

Here, the f4 and *f4* are included in each their version, in order to make the text easier to interpret, rather than using a mere repetition.

In an evolutionary context, this micro-function is overall concerned with *ordering and re-distribution of resources - nothing superfluous, nothing missing; securing effectiveness at the purpose level, while maximising homeostasis for the invironment to facilitate* [mechanisms that maximise effectiveness and thereby increase **[the capability of evolutionary potentials that enable future phylogenetic differentiation]**].

In short, in the context of evolution, the focus of this micro-function is:

*...the ordering and redistribution of resources (prioritization) (f4)*
...to maximise effectiveness (f4)
**...that indemnifies phylogenetic differentiation (F6)**

The essence of f4*(f4)* (Function 4's sub-function 4) is an accentuation of Function 4's focus (effectiveness, prioritization), which at **F6**(f4*(f4)*) happens from the perspective of a system's evolution, **F6**. Effectiveness as an activity is focused on optimising the capability and capacity for bringing about the result intended for the system – that is, securing prioritization initiatives that will safeguard the system's raison d'être.

Besides prioritization, Function 4's keyword 'cycling' is emphasised here as the mechanism of repetitive actions that keep returning to a specific point of departure from which to start again, and again, ..., – that is, an iterative and incremental strive toward a goal. The incremental nature in terms of repeated steps is a built-in trait at evolution in the sense that we know this concept. The prioritization in terms of decision-making strategies is a good candidate for such process, and which is valid for both individuals and groups of resources.

**Principle**: Survival of the fittest (combination of) decision-making strategies
- ... where the 'ordering and redistribution' entails decision-making *(f4)*
- ... where aspects of 'regulation' implicates the prioritization that enables 'effectiveness' (f4)
- ... together they provide parts of the design for maintaining 'sustainability, regeneration and evolution' necessary for management of the change that constitutes evolution **(F6)**

**Mechanism**: Natural selection of decision-making preferences
**Emergent property**: Selected phenotypes

For general systems, **F6**(f4*(f4)*)'s Principle, Mechanism and Emergent Property are:

***Principle**:* Survival of the fittest prioritisation strategy
***Mechanism**:* Natural selection of prioritization preferences
**Emergent property**: Selected resources

## *16.5* **F6**(f4*(f5)) Differentiation, Survival of the Fittest Orchestration*

From Table 2:

**Function 5** in the Mereon Matrix assures the coordination and orchestration of transformed resources to minimise risk and optimise quality.

*Keywords*: coordination; patterning; mutability; timing; multitasking; complexity.

The focus here is straightforward: "the coordination and orchestration of transformed resources to minimise risk and optimise quality", while the keywords add a couple of further aspects in focus: 'patterning' and 'timing'.

Mereon Matrix's template description for ***F6***(f4*(f5)*) (compressed):

*Coordination of quality and risk, timing and patterning to facilitate* [maximising effectiveness (i.e. ordering and redistribution of resources) that **[safeguard the coherence mandatory for regeneration, sustainability and evolution]**].

In an evolutionary context, this micro-function is concerned with the *mechanisms for coordination – while attending to fidelity, Time, timing and place – at creation, re-creation, and differentiation (patterning) in order to facilitate* [mechanisms that maximise effectiveness and thereby increase **[the capability of evolutionary potentials that enable future phylogenetic differentiation]**].

In short, in the context of evolution, the focus of this micro-function comprises is:

...*orchestration of patterning and fidelity, Time and timing (f5)*
...to maximise effectiveness (prioritization) (f4)
**...that indemnifies phylogenetic differentiation (F6)**

It is built-in traits of both Functions 4 and 6 to strive for continuous improvements (respectively 'effectiveness' and 'regeneration and sustainability'), while also Function 5 seeks to optimise the system in various ways. So, the combination of these functions designates continuous improvement, here based on Function 5's combination of activities. One of them is 'differentiation' (i.e. patterning, establishment of patterns, specialisation). Another is 'Time and timing', and the third is the attendance to 'fidelity' (quality management and risk management, the latter striving to balance advantageous changes and detrimental changes, and/or to preventing (the impact of) detrimental changes), together summarised in the concept of 'orchestration'. The option of choosing any other possibility for change than the optimal one might lead to tangential developmental paths that ultimately lead to chaotic systems rather than ordered / structured systems. In the biological example of evolution, the fitness criterion serves as a barrier toward a chaotic omnidirectional evolution.

According to Smaldino, "Group-level traits are possible when individuals display both *differentiation* and *organization*." ([73] *page 244*). By 'differentiation' he refers to individuals taking on different roles – equivalent to a process of patterning; and by 'organisation' he means differentiated individuals coordinating and cooperating on a shared purpose. This is the same meaning as the similar concepts in the template information model. Thus, the patterning in terms of speciation and specialisation are major activities in this sub-function.

**Principle**: Survival of the fittest orchestration

... where the concept of 'fittest' among others includes aspects of 'fidelity' and 'patterning' – with 'time' and 'timing' as built-in constraints *(f5)*

... where aspects of 'regulation' implicates the prioritization that enable 'effectiveness' (f4)

... together they provide parts of the design for maintaining the 'sustainability, regeneration and evolution' necessary for management of the change that constitutes evolution **(F6)**

**Mechanism**: Natural selection of the fittest orchestration
**Emergent property**: Selected phenotypes

For general systems, **F6**(f4*(f5)*)'s Principle, Mechanism and Emergent Property are:
***Principle**:* Survival of the operational opportunities that benefits the system the most
***Mechanism**:* Natural selection of the fittest orchestration
**Emergent property**: Selected resources

## *16.6* **F6**(f4*(f6)) Evolution of Evolvability*

From Table 2:

> **Function 6** in the Mereon Matrix imprints resources with what is necessary to safeguard realization, sustainability and evolution.
>
> *Keywords*: clarification; prediction; fine-tuning; coherence; condition; evolution.

The point here is evolution and the directionality that safeguards realization of the system's purpose, while ensuring long-term coherence and sustainability.

Mereon Matrix's template description for ***F6***(f4*(f6)*) (compressed):
*Warranting coherence of resources at the evolution to facilitate* [maximising effectiveness (i.e. ordering and redistribution of resources) that **[safeguard the coherence mandatory for regeneration, sustainability and evolution]**].

In an evolutionary context, this micro-function is concerned with those *mechanisms that establish evolutionary potentials within a population to safeguard regeneration, coherence and sustainability and enable evolution while maintaining invironmental homeostasis to facilitate* [mechanisms that maximise effectiveness and thereby increase **[the capability of evolutionary potentials that enable future phylogenetic differentiation]**].

In short, in the context of evolution, the focus of this micro-function is:

*...safeguarding regeneration, coherence and sustainability, while enabling evolution (f6)*

...to maximise effectiveness (prioritization) (f4)

**...that indemnifies phylogenetic differentiation (F6)**

Every resource is susceptible to the system's evolutionary mechanisms, directly and/or indirectly. As a consequence, the selected resources for each traversal of the functions exhibit gradually refined mechanisms for establishing the variance necessary for continued future evolution.

**Principle**: Survival of the fittest evolutionary capability

... where 'evolution' entails 'coherence, regeneration and sustainability' *(f6)*
... where aspects of 'regulation' implicates the prioritization that enable 'effectiveness' (f4)
... together they provide parts of the design for maintaining the 'sustainability, regeneration and evolution' necessary for management of the change that constitutes evolution **(F6)**

**Mechanism**: Natural selection among evolutionary mechanisms
**Emergent property:** Selected phenotypes exhibiting sustainable mechanisms for establishing the variance necessary for continued future evolution

For general systems, **F6**(f4*(f6)*)'s Principle, Mechanism and Emergent Property are:
***Principle**:* Survival of the fittest evolutionary capability
***Mechanism:*** Natural selection among evolutionary mechanisms
***Emergent property:*** Selected resources exhibiting sustainable mechanisms for establishing the variance necessary for continued future evolution

This corresponds to evolution of the prioritization at evolution.

### *16.7* **F6**(f4*(f7)) Integration of Phenotype within the Environment, Niche Construction*

From Table 2:

> In **Function 7**, the process of generating such potential, the Mereon Matrix is invisible, obscured by 'progeny' whose 'birth' produces energetic streams of new resources that are critical to its sustainability.
>
> *Keywords*: consciousness; evolutionary extension; integration; implementation; higher intelligence; endurance.

The focus here is on the process in connection with the generation of progeny – that is, how a cycle of the system's overall process is completed.

Mereon Matrix's template description for ***F6***(f4*(f7)*) (compressed):

*Completion of* [maximising effectiveness (i.e. ordering and redistribution of resources) that **[safeguard the coherence mandatory for regeneration, sustainability and evolution]**].

In an evolutionary context, this micro-function is concerned with *completion of the internal processes through actualisation of delivery and integration of output, interaction with progeny followed by closure of the system, insuring the internal resources for the next traversal of the seven functions to facilitate* [mechanisms that maximise effectiveness and thereby increase **[the capability of evolutionary potentials that enable future phylogenetic differentiation]**].

In short, in the context of evolution, the focus of this micro-function is:
*...actualisation of delivery and integration of output (f7)*
...to maximise effectiveness (prioritization) (f4)
**...that indemnifies phylogenetic differentiation (F6)**

The key functionality for Function 7 is the integration of output (i.e. progeny) within the systems environment; this involves interaction with the environment, an interaction that may be of a manipulative kind, in order for the resource and/or its progeny to survive and generate renewed possibilities – the next generation. In the context of evolution, this obviously leads to thinking of the 'niche construction theory'.

**Principle**: Survival of the fittest integrative capability

... where 'integration' involves exploitation of the environment *(f7)*

... where aspects of 'regulation' implicates the prioritization that enable 'effectiveness' (f4)

... together they provide parts of the design for maintaining the 'sustainability, regeneration and evolution' necessary for management of the change that constitutes evolution **(F6)**

**Mechanism**: Natural selection of integrative skills

**Emergent property** (slightly modified from [1]): Selected phenotypes integrated within their environment, each engraved with mechanisms for engaging in future evolutionary processes.

This leads to the refinement/fine-tuning of resources essential to achieve sustainability in a long-term perspective, and to secure continued survival through coherent genomic evolution.

For general systems, **F6**(f4*(f7)*)'s Principle, Mechanism and Emergent Property are:

***Principle:*** Survival of the fittest integrative capability

***Mechanism:*** Natural selection of integrative skills

***Emergent property:*** Selected resources integrated within their environment, each engraved with mechanisms for engaging in future evolutionary processes.

# 17 Assessment Against Supplementary Information from the Template Information Model

**Abstract.** The overall outcome of the extended template information model and of the Unifying Theory of Evolution is analysed against major unused properties of the Mereon Matrix.

**Keywords.** Mereon Matrix, template information model, structural aspects, spatio-temporal nexus, genesis, jitterbugging

The overall outcome of the extended template information model and of the Unifying Theory of Evolution is analysed against major unused properties of the Mereon Matrix. This approach turned out to very useful in [1] and constitutes an additional assessment effort besides the assessment points in Section 14.2.7 in order to identify potential conflicting information. The examples indicated in Section 14.2.8 are discussed in the following sections.

## *17.1 Structural Aspects*

The inclusion of the three perspectives within the structural aspects turned out to be particularly beneficial in the present modelling process and solved the deep controversies among schools of evolutionary theories. It brought a new perspective on the application domain and its theories and cleared up the messy mixture of theories. It may be necessary here to state that the author of the present paper is neither a devotee nor an opponent of either of these theories alone but rather have a pluralistic perspective on evolutionary theories. The reason is simply that the Mereon Matrix's template information model shows us again and again that a pluralistic approach likely is the accurate one; we have seen repeatedly how a choice scenario seldom is an 'either/or' but normally a 'both/and', and the outcome of the present modelling effort confirmed this.

For the purpose of understanding, it may be beneficial to look at another example from biochemical reactions within the body: the perspective of the Inevitable Evolution Theory would be analogous to chemical reactions entirely driven by thermodynamics. Then, enzymatic reactions would correspond to the perspective of the Standard Evolutionary Theory; and finally, co-enzymes and other biochemical components that co-facilitate or dampen the enzymatic reactions would correspond to the perspective of the Extended Evolutionary Synthesis in the present study. More examples might be obvious, but they have to come in later studies. At least, identification of the three axes seems to not be coincidental. The progression between the three perspectives (here 'data', 'information' and 'knowledge') worked for the Unifying Theory of Evolution, but likely will be different for other modelling efforts. It is anticipated that it will in some way depend on the function and/or on the immediately higher micro-function.

Thus, it is not feasible to express anything about the general applicability of the progression in the three perspectives.

The three perspectives is one structural aspect of the matrix. In the beginning of Part 4, it is further mentioned that the Mereon Matrix is composed of two dynamic polyhedra, an outer120-faceted (~the context) and an inner 144-faceted polyhedron (~the core), plus a handful of 2/3 trefoil knots. The two polyhedra are further mentioned in Section 17.4, to which is referred. The Trefoil knots are said in [2] to be an aspect of timing, since they are established as a function of the dynamics of the two polyhedra and follow these.

### *17.2 First Principle 0, Spatial-Temporal Nexus*

> *"The Mereon Matrix is perceived as a spatial-temporal nexus, a coherent vibratory pattern that is structurally independent of the external environment and yet inseparable from it."*

In a long-term perspective, evolution is concerned with the trajectory of a circumscribed population's traits over time and space; for instance, consider the concept of migration of man and the implications for evolution of the human traits we have today (see for instance [186] and [187]).

Although the present study is not explicitly focussed on the connection between space and time, this is obviously an integrated part of the system of evolution. The literature on theories of evolution deals extensively with this; for instance, one concept of interest is 'local' versus 'global' (cf. e.g. the concepts of 'territory' and 'habitat', i.e. **space**-related concepts). Another example is the 'evolution' as change as a function of **time**, so the concept 'spatial-temporal nexus' is at the core of evolutionary theories, even if this concept is only used explicitly by a few, such as [54] and [65]. The latter reference explicitly uses the concept 'spatiotemporal model' as the focus of modelling dynamics, where Kapeller and co-workers present results from studies on the population dynamics for his model species (the green oak leaf roller, *Tortrix viridana*) on neighbouring trees over 200 generations.

Still other examples are generational time and that the time perspective of cellular and tissue differentiation organism-wide as a function of time from the fertilized egg to senescence and death reflects the spatial-temporal nexus, the connection between space and time.

### *17.3 First Principle 1, Unity and Diversity*

> *"The Mereon Matrix is inclusive of all forms, and diversity of form and perspectives are seen in its unity."*

An analogy to the First Principle 1 for the present study topic would be that the nature of nature as we know it today represents the unity, established through the diversity of all species and their environmental conditions.

There is no doubt that the unity in nature arises through the diversity, and that this remarkable balance is based on interspecies cooperation, interferences and interactions, where each individual and each species is one piece of an incredible puzzle.

## 17.4 First Principle 2, Context and Core

*"The Mereon Matrix has a context with the capacity to set orientation, determine polarity and direct functions, and a logical core that is responsible for maintaining sustainability and ordering functionality."*

What immediately comes to mind is the individual versus its social context (as discussed for instance in Sections 5 and 6). It is demonstrated that the individual (~the core) behaves differently when it is on its own compared to its behaviour within a group context (~the context), or as Webster and Ward express it "... a complex and context–dependent influence of sociality upon individual behaviour." ([66] *page 765*). That is, the context "determines polarity and directs functions", while the core is the one "responsible for maintaining sustainability and ordering functionality". In a Mereon Matrix context, a system includes both, and both are necessary for sustainability and viability of the system.

## 17.5 First Principle 3, Possibilities

*"The Mereon Matrix requires that the local environment provides adequate resources, 'possibilities' that are simplest and stable."*

This was concluded fulfilled for every function at the macro and micro level in the pilot study. Each successive emergent property constitutes 'selected resources' of various kinds, so inevitably the local environment/invironment provides resources for successive functions. Should one function fail to process the resources – or if processing is irrelevant given the conditions – then that part of its output will be identical to its input, and this may still serve as input for the next function. Specifically for the system of biological evolution, the possibilities comprise the metabolic state and system of the organism, therein the pool of genetic variation.

In case of inappropriate resources, the system will either come to a halt and decay (too few or inappropriate resources, leading to extinction), or burn-out (too many resources, leading to existential fight for survival).

## 17.6 The Genesis

The Genesis of a Mereon Matrix is the process that leads from a regenerative seed to a full matrix system (see Chapter 8 in [2]). At this point, it is worth pointing out that the progeny of the Mereon Matrix (at Function 7) comprises 10 dodecahedral 'daughters'.

An example with similarity to the genesis, highly illustrative yet not unequivocally drawn from the domain of evolution, is the process of fertilizing an ovum followed by the subsequent processes leading to the zygote, and on.

The final outcome of the genesis – in the context of biological evolution – is a biological system. This leads to the discussion of pre-life evolution in Chapter 11. We have a fractal template, and thus, the fractal principle has to be visible from the smallest to the largest entity observable. Therefore, the question arises "where to find the pre-life evolution in the Unifying Theory of Evolution?" The answer is suggested to be "the next fractal level of F6(f4(f1)) of the perspective Inevitable Evolution Theory and 'down'-wards toward increasing levels of detail".

### *17.7 Jitterbugging*

A jitterbug denotes a geometric 'dance' representing the dynamic transformation from cubic/octahedral symmetry to dodecahedral/icosahedral symmetry, with a twisting-expanding/contracting, inside-out movement. There will be a similar jitterbugging – whether or not we are able to identify it – at the present or rather next level of detail.

Jitterbugs are determined by the dynamics of the Mereon Matrix's 120 polyhedron; see [2]. The magic of the Mereon Matrix is that the dynamics of the 120-polyhedron defines the motion of 5 jitterbugs (corresponding to the number of each of the mentioned polyhedra).

A single jitterbug starts with an octahedron (corresponding to F3, here: F6(f4(f3))) and transforms to the cube-octahedron (Function 1, here: F6(f4(f1))), and in this process defines an icosahedron (Function 2, here: F6(f4(f2))), and vice versa. In an information model interpretation, this corresponds to the dynamics that tie Function 1, 2 and 3 together. Following the establishment of the jitterbug, this also embraces the dodecahedron (corresponding to F6(f4(f4))), the rhombic polyhedra (corresponding to the combination of F6(f4(f5)) and F6(f4(f6))) and finally one scaled, large icosahedron (corresponding to F6(f4(f7))) that hides everything inside it.

This explains why there is an invisible step (transition) between the basic principles and the elaborate ones: Looking at the information contents of functions F6(f4(f1)), F6(f4(f2)) and F6(f4(f3)), one sees the basics of the three evolutionary theories and that there is a natural progression from the simple basic selection principle behind evolution at F6(f4(f1)) to the selection mechanism at an individual level and on to the selection mechanism at a group level. In comparison, the Functions 4 till 7 (i.e. F6(f4(f4)) .. F6(f4(f6))) have a more abstract nature: decision-making, differentiation, and evolvability, while integration in F6(f4(f7)) has a icosahedral basic, practical nature like F6(f4(f3)). See the illustration in Figure 4 in Section 18.3.1.

### *17.8 Input, Throughput, and Output (Chapter 6, page 105 in [2])*

The three axes in the 120-polyhedron of the Mereon Matrix constitute different symmetry axes in the geometry and dynamics, respectively the three fold (~input), four-fold (~throughput) and five-fold (output) axis. The correlation between the symmetry axes and the three types of functions are related to their behaviour in the dynamics: the vertices of the input axis opens for inwards flow, the vertices of the throughput axis remain closed, and the vertices of the output axis opens for outward flow. Comparing these three axes with the three perspectives in the Unifying Theory of Evolution suggests that:

a) The Inevitable Evolution Theory compares to the input axis. The openings for this axis are very small, and hence by analogy this might compare to the basic and simple input mechanism behind all biological, biochemical and physical processes: quantum mechanics, and in particular energy states (thermodynamics).
b) The Standard Evolutionary Theory compares to the throughput axis, corresponding to mutations (in the widest sense: changes in the genome) as the processing mechanism behind evolution.
c) The Evolutionary Extension Synthesis theory compares to the output axis, corresponding to integrative processes establishing the phenotype through

developmental processes that includes interaction with the environment. The phenotype is the visible measure toward the external world of that which has taken place in the regulatory developmental process from genome to phenotype, including the effect of plasticity. Still, even if the mechanism behind the Extended Evolutionary Synthesis is epigenetics based, in a higher view it is nevertheless changes in the genome and its derived products that implement plasticity.

## *17.9 Quantitative Aspects*

What astonished the author incredibly in the similar assessment effort in [1] (*page 513*) was the following: "The predominant form of DNA, the B-form, has 10.5 base pairs per turn; the Z-form has 12 base pairs per turn; and, the A-form has 11 base pairs per turn. These data coincide with the number of 90° angles in the Mereon Trefoil, as seen from Chapter 10", namely respectively 10.6, 12 and 11.

No quantitative aspects are included in the present modelling effort. Still, a single quantitative aspect pops up; however, one needs to be extraordinarily careful here in order to not give rise to a bias. In the Mereon Matrix the various polyhedra appear in different quantities within the 120-polyhedron: 10 tetrahedra in five matched pairs (corresponding to various types of genomic changes), five cubes (corresponding to Function 1), five icosahedra (~F2), five octahedra (~F3), one dodecahedron (~F4), five rhombic dodecahedra (~F5), one rhombic triacontahedron (~F6), and one scaled icosahedron (~F7). One cannot draw a full analogy, for the following reasons: 1) for instance, niche construction is paralleled in the three component perspectives of the unifying model; and 2) the domain knowledge is incomplete. Still, looking apart from these hesitations, one does notice that there seems to be an analogy in Table 1.

Moreover, there are a multitude of types of genomic changes corresponding to a Function zero, the tetrahedron, - that is, the system's input resources: micro-functions 1 .. 3 and 5 have parallel sub-topics at the highest bullet level), while for instance micro-functions 4, 6 and 7 has but a single sub-topic for each perspective – but of course this also depends on the author's hierarchical grouping and merging of topics, and thus depends on the level of detail in the literature on the topics addressed for given functions.

This kind of anticipation regarding quantitative features is guesswork, until more experience has been gained with modelling according to the template information model at the present level of detail. If the hypothesis is correct, then it shows that there will be more topics of relevance to include in the Unifying Theory of Evolution (see for instance Function 3 under the Inevitable Evolution Theory).

Following this, it should be mentioned that Koonin mentions power laws in [9] (*pages 1025 and 1021,* see also Section 18.1) in connection with gene duplication where the distribution of family size in all sequenced genomes follows a power-law like distribution. Koonin also mentions another quantitative aspect in connection with regulatory genes and regulatory overhead that shows an (nearly) quadratic scaling ([9] *page 1023*). Important is a conclusive remark of his: "The differential scaling of functional genes with genome size … suggests the existence of an entire set of fundamental constants of evolution. The ratios of the gene duplication rates to gene elimination rates that determine the exponents of the power laws for each class of genes appear to be the same for all tested prokaryotes and invariant with respect to time … (*refs*)." ([9] *page 1025*).

The present author's best guess would be that the quantitative aspects may appear in the next micro-level and in full in particular in function F3(f3(f3)) (See Table 3).

### 17.10 Breathing and Birthing

'Birthing' is an integrated part of Function 7 and a kernel aspect of all evolutionary theories, namely reproduction, while in the present study of the 2nd micro-level, the focus in Function 7 is on the integrative aspect.

'Breathing' is the dynamics of the two polyhedra that together constitute the Mereon Matrix. In the dynamics, the 120 polyhedron expands to a 180 polyhedron and back to the 120 polyhedron. Similarly, the 144 polyhedron expands to a 300 polyhedron and then back to the 144 polyhedron. Breathing is paralleled to the mechanism that drives the progression from one function to the next until the end of a traversal, through the succession of functions, and then continuing with the first function again.

One analogy that might be considered as an analogy to the breathing of the matrix is the repeated cycles of accumulation of cryptic variation (i.e. an expansion in Mereon Matrix terms) that at some point becomes advantageous and therefore will be exploited (in the new phenotype) and then followed by a 'release' process of genetic accommodation (i.e. a contraction in Mereon Matrix terms) – thus reducing the amount of cryptic variation.

### 17.11 In Summary

The pilot study showed that the entire domain of human molecular genetics can be covered in one model by usage of features in the Mereon Matrix, [1]. The present study operates at one fractal lower level and in a specific niche; and again the modelling efforts identified the majority of the Mereon Matrix features, even stronger than for the pilot study. All the major properties of the Mereon Matrix seem to have at least weak analogues in the Unifying Theory, and one should not expect more at this point.

# 18 Reflecting on Validity

**Abstract.** This chapter is concerned with assessment of the outcome in the broadest sense - that is, it includes an analysis of the validity of the template information model, of the risk of bias, and of the Unifying Theory itself.

**Keywords.** Unifying Theory of Evolution, template information model, validity, self-assessment, bias, General Systems Theory

## *18.1 Validity of the Template Information Model*

What is more needed before it can be concluded that the Mereon Matrix template information model is valid as a universal template for modelling a knowledge domain? Were there any flaws in the template information model?

The main problem in demonstrating the validity and generality of the template information model is that the study itself constitutes constructive assessment in that it was necessary to extend the template model with the description of an additional fractal level while using the outcome of that developmental process. There is a huge difference between constructive (also called formative) and summative evaluation. Constructive evaluation comprises evaluation activities that are completely intertwined with the development activities throughout the project (or for a circumscribed period/phase), while summative evaluation is concerned with evaluation at an end point in a developmental path or phase. The difficulty at concluding on validity in a scenario with constructive evaluation is the lack of a frame of reference against which to assess and in particular because every decision making influences the solution space for subsequent decisions; see a little more detail in [188]. Consequently, one has to be extra careful about biases: a method or theory may indeed be highly successful in application, but one cannot fully conclude on the method's generality and validity in terms of various kinds of validity, until assessed independently in another setting.

As concluded in Section 18.2, there is a risk of circular inference bias if it is unconditionally concluded that the template model is valid for the present modelling purpose. Subjectively assessed, the template information model has been extremely beneficial; without it the present study would not have been feasible, in particular given that the author had little previous insight into the domain of biological evolution.

No flaws were encountered in the template information model. The 2$^{nd}$ micro-level function descriptions were developed iteratively as part of the present modelling efforts and gave no cause for concern. A few more keywords for the functions might make it easier to interpret these and match them with concepts in various knowledge domains; in particular 'efficiency' and 'effectiveness' are two terms that are missed especially for the medical and biological domains; however, and unfortunately, they are also terms that have various meaning in various domains and therefore not suited as keywords. Moreover, when one genuinely understands the Mereon Matrix functions, such additional terms are no longer necessary.

We suggest that the generality and validity of the Mereon Matrix's template information model may be judged and hence documented as valid (at least for certain modelling tasks) when all the significant characteristics of the Mereon Matrix have shown particularly beneficial in at least one modelling effort, like the three perspectives in the present study, and the 7 functions both here and in the pilot study. With 49 subtopics in molecular genetics there will probably be chances of trying out all aspects of the template model within this single knowledge domain.

Within the pilot study, we identified one astonishing quantitative aspect of alignment with the template information model; see Section 17.9. Here, a glimpse of the quantitative characteristics has been observed in terms of the relative numbers of topics in Table 1 and a couple of observations in the literature, but nothing that compares to evidence. It would be nice to find a full-blown example of the scale-invariant properties, i.e. the power laws, as for instance explained together with the fractal nature of nature in [190]. Koonin refers to the power law in [9]: "many genes belong to large families of paralogs which form a characteristic power law distribution of the number of members (*refs*);" ([9] *page 1021*), so this regularity of systems indeed has already been identified within the domain of evolution, but the present study has not been elaborated to a level of detail where quantitative aspects are obvious to dig into.

In summary, what is further needed are studies, where each of these template properties turn out to be drivers of the modelling work, similar to what happened in present study with the three perspectives.

### *18.2 The Risk of Bias*

The number of potential biases is huge and naturally depends on the nature of the study. They differ from social and psychological studies to medical studies and biological studies and on to for instance systems development studies. A review of biases encompassing various types from these disciplines may be found in [15].

#### *18.2.1 Local Minima Bias*

In large development projects there tend to be a succession of decision-making points; every interpretation of the template information model constitutes such succession of decision-points. The risk of a bias arises when the decision-making constitutes one out of a series, and when the decision-making basis in a given situation (unintentionally) points at a solution that constitutes a local optimum. In such a scenario, each decision inevitably influences the solution space for subsequent decisions. That is, although a decision appears optimal within the specific context, it may not be so in a larger perspective. This is somewhat similar to the problem of accuracy discussed for quorum sensing in [101], see Section 8.1.3.

In the context of the present modelling task, the fact that the function descriptions are elicited for the 2$^{nd}$ micro-level while using the template model from [2], which only includes the macro and the 1$^{st}$ micro-level functionality, makes this type of bias particularly relevant. The approach for handling this bias is in the iterative and incremental approach while enforcing details within the validation step (Step 7). If the pilot study model were wrong then continued modelling of the next fractal level would not succeed, but may succeed for modelling of a particular single micro-function on its own.

### *18.2.2 Judgemental Bias*

There are various judgemental biases at which interfering psychological factors affects the outcome of an assessment process, for instance previous experience of the performer or non-blinded events; see [15] or [191]. The author is aware that there is a risk of something similar to this bias, corresponding to the old saying that "when one looks for a hammer, everything looks like it could be used for a hammer". Or as an editorial in Nature says "The human brain's habit of finding what it wants to find is a key problem for research." ([192] *page 163*). They call this a 'cognitive bias'. Moreover, they say "One enemy of robust science is our humanity – our appetite for being right, and our tendency to find patterns in noise, to see supporting evidence for what we already believe is true, and to ignore the facts that do not fit." ([192] *page 163*). These aspects are part of 'judgemental bias', but there is also more to this bias, such as a number of variants concerning people's ability to judge probabilities of events, situational effects (e.g. experimenter's experience and expectations), and more.

The Mereon Matrix has served as the template for the functionality looked for, so it is obvious that the author's personal and professional perspective may unconsciously and unintentionally be biased; thus, there is a risk that one may identify characteristics that one already knows and may overlook others.

The author's professional background and role within the context of this work is outlined in the Foreword. This should warrant a fair degree of objectivity in the study despite the qualitative nature (and the consequential degree of subjectivity), because there has been no personal interest to pursue or promote specific evolutionary theories and/or downplay others.

There exists no measuring stick with which to compensate for the risk of judgemental bias or to formally assess the validity of the model in absolute terms, beyond the evaluation approach suggested in Section 2.2.7. Apart from this, the author has continuously self-reflected on the process of the study to understand how personal values and views influenced the decisions-making process, as Jootun and co-workers recommends ([193]): 'Reflexivity' is "the continued process of reflection by the researcher on his or her values, preconceptions, behaviour or presence and those of the participants, which can affect the interpretation of responses." ([193] *page 42*). In this respect, it has definitely been an advantage that the author is an outsider, because there were/are no personal relations or dependency with any of the research teams, their members or their appreciated theories within the knowledge domain, or competitive access to research grants. This makes impartiality and objectivity relatively easy, and hence, also easy to avoid becoming enmeshed with specific viewpoints.

### *18.2.3 Hypothesis Fixation*

"Hypothesis fixation occurs when an investigator persists in maintaining a hypothesis that has conclusively been demonstrated to be false (*ref*). When a specific hypothesis is addressed (rather than an explorative investigation), this hypothesis should be explicitly stated and shown as an integrated part of the assessment protocol." ([15] *page 307*).

The present study is an extension of a previous pilot study, [1], which we concluded to be successful, yet without a frame of reference for objective validation purposes, since it was a) the first such effort, and b) performed as constructive evaluation. Unfortunately, like developers in a systems development project can have difficulty at being impartial, [15], this bias is a potential risk when one is seriously engaged in a specific method (corresponds to the 'IKEA effect' in the editorial of [192]). The author is very well aware of this and has done my best to stick to a rigorous scientific approach that will compensate. At present no independent research group has repeated the efforts, and this is not likely to happen until the template information model is sufficiently known from published studies.

Like the pilot study, also the present study constitutes a constructive evaluation effort, since the Mereon Matrix template information model is elaborated with an additional fractal level prior to its application. There is still no availability of a frame of reference – because it is an explorative study that at the same time generates the future frame of reference. Consequently, beyond the validation in the Method section the validity resides in the author's professional and scientific competence.

There are a couple of examples, where the template model overrules the dominating perception in the literature, and thereby they may become focus points in a discussion of the risk of a hypothesis fixation bias:

- The pluralistic model that unites competitive theories within the literature rather than solving the controversies in a traditional way. In the observed cases, the competitive theories ended up in each their perspective, and consequently there is no conflict in the unifying theory with either component theory.
- Placement of 'niche construction' in the model. Laland and co-workers correlates niche construction with 'developmental biology' (Function 1), while in the Unifying Theory it is referred to Function 7 and distributed on more than one perspective. Nevertheless, there is no conflict with the template model and no implied consequence for the niche construction theory, just diverging opinions/verbal arguments.

In summary, it is at present too early to consider the risk of a hypothesis fixation bias, since there is no conclusion of falseness that might give a hint, but the author has had a focus on this risk.

### *18.2.4 Circular Inference*

"Circular inference arises when one develops a method, a framework, or a technique dedicated to a specific (population of) case(s) and applies it on the very same case(s) for verification purposes." ([15] *page 265*). The present study constitutes a constructive evaluation effort, so this bias is an obvious candidate. Moreover, in principle it is not possible to objectively judge the accuracy of the outcome of one's own modelling efforts, unless applying method triangulation, which is not feasible for the present study, or await the outcome of other research teams repeating our studies. However, the author is aware of this risk and takes precautions accordingly.

An aspect of the study of particular relevance to discuss in this context is the extension of the template model followed by application on the same object of study. There is no feasible alternative methodology for the study at this point of the global endeavour. The approach here was to prepare the extension prior to the application – and this was accomplished in a first complete version for all 343 functions at the 2$^{nd}$ micro-level prior to even deciding on the choice of evolution as the topic of a first exploration at the 2$^{nd}$ micro-level. However, refinements were necessary during the modelling work to compensate for ambiguity, comprehensibility or lack of coherence – after all it is a constructive evaluation effort. There were no significant changes in the formulation of second micro-level functions during the application and zero changes in the existing function descriptions for the macro- and the first micro-level functions in [2].

In conclusion, it is anticipated that the impact of this bias on the outcome of the present study is minor. To ultimately conclude on the general applicability of the template information model, one has to rely on other researchers performing similar modelling efforts with the same template information model. That may take years since new models have to start at the macro level and also because it is anticipated that few studies will need modelling at the 2$^{nd}$ micro-level to achieve the desired level of detail for a particular purpose.

### *18.2.5 Inclusion Bias*

There is a risk of an inclusion bias since only a proportion of review authors and other sources of information used here perform systematic reviews, nor is the present study performing systematic reviews – that is, in neither case the literature knowledge base is exhausted.

An experience was that the exhaustive, unbiased reviews are scarcer than within the knowledge domain of human molecular genetics in general, presumably because of the complexity and the difficulty at performing experimental studies. Further, the tendency of journals to limit the space available to papers is counterproductive in this respect as regards the synthesis of objective and exhaustive reviews. Still, all of the reviews are immensely valuable when the reader is aware of such conditions and takes the necessary precautions at the interpretation.

One may say that a weakness of the present study is that it was necessary to identify a consensus without performing an exhaustive systematic review, simply because systematic reviews require efforts way beyond what is justified for the present study purpose. Further, how is a consensus achievable when the knowledge base repeatedly shows conflicting viewpoints and exclusivity? Where there are controversies within the literature, the author has done her best to incrementally build up the evidence supporting the model based on complementary papers within the literature that were judged to be convincing. The value norm was to achieve a coherent wholeness while looking for trustworthiness of the papers included (scientific quality and coverage, unbiased), and using the literature as the source of evidence while following own prescriptions for quality of input.

The symptom of 'hiding results to seek the truth' as discussed by MacCoun and Perlmutter in [194] (also called 'asymmetrical attention' in the Nature commentary of [192]) is obvious in the literature on evolution. Controversies were experienced where (sets of) research groups with strongly opposing views tend to not always objectively reference each other's research and conclusion, thereby border lining opinion papers rather than traditional reviews in the strict scientific sense. This brought the attention on a variant of inclusion bias: when a systematic review is not feasible or when the necessary efforts for exhausting the literature's resources are not justified for achieving the study purpose, and/or when the knowledge domain includes biased / focused articles and reviews, then one's own literature search unintentionally inherits a risk of an inclusion (or information) bias. This may be further enforced when using reference lists as one's source of information to trace relevant articles, as suggested in Section 2.3.

### 18.2.6 In Summary

When carefully examining the nature of the present study, virtually all of the biases discussed are built-in, candidate biases at risk. However, these biases are well-known to the author and have been in focus throughout the study.

The uncertainty generated by bias, and hence the impact on the validity of the modelling outcome, subjectively assessed, may reveal itself in terms of the (lack of) smoothness, completeness, continuity and coherence of the final model. The walk-through the sub-functions in the sequence $F_x(f_y) \rightarrow F_x(f_y(f_1))$ .. $F_x(f_y(f_7)) \rightarrow F_x(f_y+1(f_1))$, and so on - for all x and y – raised no concern as regards appropriate coherence, continuity and completeness.

The discussion of the relevant biases has not exposed any serious risk of bias for the present work. The 'local minima bias' is probably compensated for by the iterative and incremental approach, where previous decisions were constantly revisited and re-iterated. The 'judgemental bias' is anticipated minimised by continuous reflexivity and by the outsider position freeing the author of entanglement to existing and competitive theories. The timing is immature for judging the risk of a 'hypothesis fixation bias'. There is a particular risk of a 'circular inference bias', because the applied level of the template information model was constructed and then used, which means that one has to be careful at concluding on the validity of the template model. This study is clearly the victim of and therefore inherit the 'inclusion bias' since exhaustive systematic reviews of the various topics are not available in the literature. Nevertheless, all of this is anticipated to have inconsequential impact, when considering that the principal issue is that of populating the individual cells of the Unifying Theory of Evolution with trustworthy sub-theories and information that all together make the entirety make sense.

In conclusion, there seem to be no sign of serious impact from the discussed biases on the outcome of the present study.

## 18.3 Validity of the Unifying Theory of Evolution

The main problem in demonstrating the validity of the Unifying Theory is the lack of a frame of reference against which to assess the validity, whether by statistical means or otherwise. Therefore, one has to look at various indications of validity or lack thereof.

### 18.3.1 Indicators of Validity

For a qualitative study like the present, it is not straightforward to apply conventional measures of validity, but one can apply analogies once a concept is understood. Thus, the aspects of validity of the Unifying Theory of Evolution to consider include for instance (as mentioned in Section 14.2.7, bullet 8):

1) Construct validity – that is, does one really "measure" (in the present context: "capture") that which one believes or intends: this issue is addressed above in the discussion on potential biases, to which is referred;
2) Internal validity (degree of compliance between the perceived meaning and the reality, i.e. with minimal bias): this is related to the ability of interpreting

the template model in the context of a particular knowledge domain. Also this is addressed in the discussion on potential biases, to which is referred;

3) External validity (generalizability to other contexts of investigation): The template information model is in the process of being applied in practise in other settings, like for a classroom management model and for business consultancy; the former model is highly successful, the latter activity is only in its infancy. Neither of these applications goes beyond the macro-level of the functions. Also the model of the origin of life in [184] operates at the macro level. Moreover, neither of these so far has been formally evaluated. However, the application range and hence the generalizability makes the external validity promising;
4) Empirical validity (accuracy toward the true value of a measurement, or – in an evolution context – the true meaning expressed in the literature included): It is easy to bias the message of a paper, when one boils the meaning down to picking a sentence or bringing the message in another context. For this reason, the author has been cautious at securing the context and content of the authors' statements, and in a quite a number of cases bring the meaning as a citation to give the full credit to these authors;
5) Rational validity (coverage or representativeness of characteristics): The template model is fairly stringent, so the coverage and representativeness are a matter of putting more or less details on the individual micro-functions, and hence, this will depend on the maturity of the knowledge domain. In the Unifying Theory of Evolution there are areas that are not fully elaborated, simply because of lack of material in the literature. This may be because it does not exist, or at least it has not been found. Instead, at least small examples from the literature are included of what is supposed to be in those parts of the model – hence, areas for suggested future research;
6) Reliability (consistent outcome): If someone else than the present author – with the same level of competence and background – repeat the present modelling study, would the outcome be different? Provided that a similar effort is invested, the answer will be "No! – likely not, but the examples would be different and so would the list of reference included be".

It seems that these aspects of validity of the Unifying Theory of Evolution are rated as fairly assuring.

In hindsight on the Unifying Theory of Evolution, we are looking for coherence, smoothness, continuity, consistency, while observing and abstracting the resulting model to find a logical flow of the diversity within the unity of the wholeness. The entirety of aspects of evaluation of the Unifying Theory of Evolution (see Step 7: Evaluation of the Model in Section 14.2) has been evaluated, partly as part of the iterative modelling process and partly in a final and dedicated activity.

That the coherence and continuity of the model is fulfilled, may be illustrated by abstracting the information in Table 1, see Figure 4.

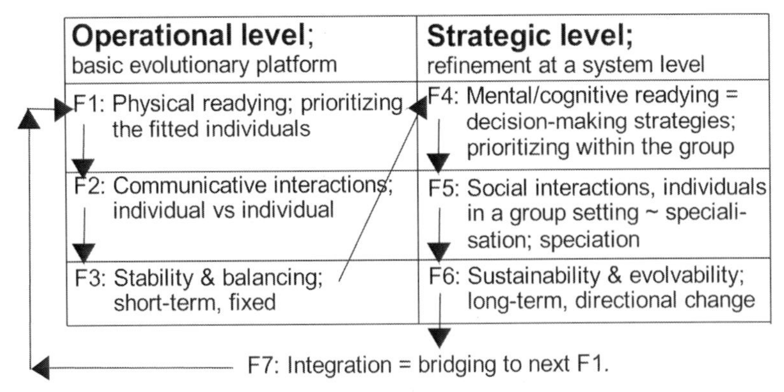


**Figure 4:** Overview of relations between the functions at an operational level versus a strategic system level.

The left column constitutes the foundational evolutionary platform in that there is little explicit objectives-orientation in the functions *per se*. For instance, Function 3 (Stabilisation) is concerned with balancing 'what is' and efficiency (doing things right), but this is short-term and focused on local goal-orientation of maintaining homeostasis within the system. Each of the functions in the right column adds in a sequential manner to the entirety of the basic platform, and yet pair-wise each has similarities with the respective counterpart function at the operational level: physical readying versus cognitive readying/prioritization, communicative interactions versus social interactions, and stability versus sustainability. And at last, Function 7 bridges to the next Function 1 by integration and implementation within the environment.

### *18.3.2 Mastering the Template Information Model*

The critical issue regarding the validity of the outcome model when applying the template information model is the understanding of the Mereon Matrix; this is not mastered or learned in a lunchbreak, like my professors at the university said about learning a new programming language or a new diagramming technique. However, the template information model is fairly stringent, so the validity has a high degree of certainty, once the matrix is understood. So, this is not the issue for the present application, but may be for new applicants using it. This is why there has been put that much emphasis into the explanation of the modelling process in Chapter 16.

### *18.3.3 Confounding Factors in the Resulting Model*

Is the model still valid if man manipulates in a way that obstructs the survival principle or the natural selection mechanisms? – like at domestication of animals, gene manipulation of crops and cloning of animals. Things happen of course, but this will not change the Unifying Theory or any of its component theories: still, only the fitted will survive, and the individuals will adapt in an attempt to survive (Function 1). Population dynamics, migration patterns, as well as group decision making will be affected for animals in captivity. Such manipulation may change the fitness landscape in more or less every respect for the involved individuals, groups and their species; thus, the selection mechanisms are partly artificial for as long as the manipulation is active. Domestication of animals may over time allow non-selected traits to vanish or flourish and may allow cryptic variation to surface in the phenotype; and it may over time alter the space for accumulation of mutations in these respects, simply because individuals that would normally be deselected may artificially be selected because they have other traits that are preferred by the manipulator. In agreement with the principle of the Mereon Matrix as a whole, the functions remain unchangeable while the resources (i.e. individuals and groups of individuals within a population) are indeed changeable. The Unifying Theory of Evolution remains valid in its entirety, and both the principles and mechanisms remain active, while the processed resources are artificially manipulated in addition to the natural processes. For as long as the manipulation is active the artificial selection will supplement and/or in part replace the natural selection if sufficiently strong. However, for instance pecking order, and other competitive principles, remain active among our domesticated animals until such aggressive traits in the population have been deselected. The interference with natural evolutionary mechanisms may indeed locally affect mechanisms like the natural kin selection, group selection, niche construction and more, but globally the wholeness of the Unifying Theory of Evolution will remain valid and active.

### *18.3.4 The Chicken & Egg Problem Explained*

There is a chicken-and-egg problem: decision-making strategies are casually applied in Section 6.3.1 – that is, before decision-making strategies are introduced in the model in Chapter 7. As said earlier, this is allowed – with caution – because of the incremental development in successive cycles of Functions 1 to 7. The implication is that decision-making strategies (and hence cognition) must have developed before the conscious teaming strategies. This seems plausible; however, the author has no means for verification of this assumption.

### *18.3.5 The Literature's Lack of Insight into Actual Mechanisms*

In many cases, the literature describes numerous variables and documents their role as factors in evolution, however, without describing actual mechanisms. To extend or extrapolate the model from this state would be speculative and therefore has been avoided. This does not render the Unifying Model invalid, only incomplete, but this is a given condition in frontier science. As these holes in the domain knowledge gradually are revealed they may further fill the model, while potentially some of the referenced studies may have to switch from one perspective to another, for instance when one learns that a mechanism previously thought to be genetic changes turns out to be based on principles of plastic adaptation. However, this may move individual pieces of knowledge horizontally in the model but not vertically from one function to another.

### *18.3.6 Exploitation of References*

Some places in the Unifying Theory of Evolution are weaker populated than others, and still when reading the rest of the book one may find small pieces of information that could also provide evidence in other places. Little effort has been invested in including such pieces at both places, when there are already illustrative examples, simply because the information gain is small.

### *18.3.7 What about the Author's (Lacking) Prior Insight into Evolution?*

The Foreword poses the question "why publish a book on evolution, when it is not the author's core competence?" Of course, the book would have been different had it been done by someone with decades of experience in evolution, while also spanning and acknowledging the existence of different perspectives. The author's unconditional advantage in this respect is the impartiality, the independence of professional schools within the knowledge domain.

As said previously, the purpose of the study originally was to assess the general applicability of the template information model as an extension to the pilot study, where the author was fully confident with the knowledge base at hand – at least at the level of detail dealt with. This study goes one level deeper in detail with a specific niche from the pilot study, thereby addressing the micro-micro level (i.e. the 2nd fractal level beneath the macro-level). So, at what level of detail will it require dedicated and deep prior domain insight into the topic dealt with in order to be able to handle the modelling with confidence? Two years of full time researching the topic of evolution has enabled the author to get an overview of the knowledge base that suffices for writing the present book. The author's guess is that "maybe already for the 3rd fractal level beneath the macro-level – at least for some niches" it will be necessary with prior in-depth knowledge for detailed modelling of a knowledge domain.

For the author, it has been a process stretched over two years, so the advanced professional from the evolution domain may identify some statements within the discussion of his/her own theories, which are a bit naïve or expressed slightly inaccurate – apologies for that. What has been important for the author is the accuracy in placement of the various small and large theories in the right place in the model, and here, the author feels confident about the end result. It is anticipated that the end result, the Unifying Theory of Evolution, would not have been any different with respect to the structure and main theories included if another year had been added.

### *18.3.8 What about That, Which is Still Unknown?*

About 1.5 % of DNA accounts for the amount of DNA that is transcribed and translated into protein (when the introns are excluded, and about 30 % when the introns are included in the count ([195] *page 952*). Apart from that a big portion is accounted for by RNAs and micro-RNAs. The latter have a regulatory role in the metabolisms and thereby also in evolutionary process; this could for instance be a part of the mechanisms behind plasticity. We know about the mitochondria and the hypothesis of their endosymbiotic origin, see among others the textbook of [195]. Further, Pavlicev and co-workers report that transposable elements "have entered the genome in past viral invasions and comprise 50% or more of the human genome (*ref*). Although their transposing activity is often suppressed, the importance of these genomic elements in introducing genetic variation, enhancing plastic environmental responses, and in particular in long-term diversification of plants and animals, is well recognized (*refs*)." ([196] *page 1082*). Moreover, "Numerous examples of co-option of TEs into a series of crucial functions have been documented, including recombination, splicing, exonification, and various modes of gene regulation (e.g., *ref*). Among the most prominent examples are effects on the regulation of adjacent genes." ([196] *pages 1082-3*). All of this suggests that transposable elements have a significant contribution to the evolutionary processes. Yet, exploring the function of every little detail of the genome and epigenome is still in its early stages.

In conclusion, there may be many and/or major surprises to be revealed some time in the future. Then, what would such revelations imply for the Unifying Theory of Evolution? Transposable elements have been proposed to increase evolvability because such chunks of genetic material are more likely to have functional consequences on the phenotype than single mutations will have (see the discussion in [196]), even when these are accumulated over time. Thus, big chunks of genetic material being inserted may lead to giant steps in the evolution rather than the small steps accounted for by regular mutational accumulation – even in the widest sense of intraspecific changes in the genome. However, this will not change the structure of the Unifying Theory of Evolution with its three perspectives and the seven functions; the genome is still a genome and will serve within the cell in the regular metabolic processes, but of course the effect on the phenotype may be drastic. And similarly, regulatory elements will still regulate the metabolic processes of the cell, individually or tissue-specific, and depending on the timing of the appearance within the life-cycle of that organism or cell, and thereby add to the aspect of plasticity and its implication throughout the model, and the effect on the phenotype may be tremendous; however, an additional perspective is not foreseen. Even the appearance of new independent genomic elements within the cell such as new transposon elements, in the same way as the mitochondria appeared, are not foreseen to change the structure of the Unifying Theory of Evolution, but large pieces of information may be added in various places. That is, a parallel line of mechanisms for evolutionary developments (i.e. a new perspective) is not foreseen. However, of course such anticipations have to be revisited in case knowledge in this respect is revealed.

## *18.4 Opportunities for a General Systems Theory?*

The present study has made it probable that modelling by means of the Mereon Matrix may be a workable approach for achieving the General Systems Theory. What is more needed for demonstrating this convincingly?

From Chapter 16, one may get a glimpse of a general systems theory, comprising the function descriptions together with the Principles and Mechanisms, transformed into an abstracted version liberated from the ties to a single knowledge domain – however, only for one out of the 49 functions at the 1$^{st}$ micro-level.

Completing the ongoing modelling for all of the 49 micro-functions would be interesting from the perspective of molecular genetics as a knowledge domain, but will delay the venture of achieving a General Systems Theory. It is anticipated that the achievement of further modelling experience with the Mereon Matrix template information model may enable one to extrapolate by analogy the abstract, generalised version(s) from existing modelling efforts, while iterating Steps 1 and 3, for instance, based on the complete domain model in [1]. Since only a subset of features of the information in the template information model has been applied, further needed is experience with each of the supplementary information in the template information model of the Mereon Matrix. This needs not be in a single modelling exercise, but may take place for the individual pieces of information one-by-one. For instance, beneficial would be the achievement of modelling the quantitative aspects in a convincing manner, as that may lead to the power functions; however, the domain of molecular genetics may not be the best choice for such purpose.

What was surprising to the author during the present study is that the laws of physics in general entered the scene at this 'early' point in the modelling (see Chapter 11), namely at the fourth level (the macro level plus three micro-levels). It was foreseen that the laws of physics would appear at the 'bottom' level for the modelling of molecular genetics. So, if the hypothesis in Chapter 11 is correct, then the complete General Systems Theory will not contain as many levels as anticipated. This again will make it more plausible that the above suggested approach will provide a good foundation for a General Systems Theory.

## *18.5 Postscript*

The conclusion is that the Mereon Matrix's template information model enabled the generation of a Unifying Theory of Evolution that integrates Darwin's evolutionary theory with recent evolutionary theories.

A retrospective view on the Unifying Theory of Evolution leads to the following three questions: 1) How to look at the existing literature on evolution, given the new unifying theory? 2) Are there other theories or frameworks that unify evolutionary theories? And 3) what's the use of the Unifying Theory of Evolution – that is, what will be its impact and potential for future research on evolution?

### *18.5.1 Back to the Essential Question*

Are there other evolutionary mechanisms than 'natural selection'?

A small thought experiment: Function F6(f4) has 'natural selection' (adopted from Darwin) as its mechanism, and the function F6(f4(f2)) .. F6(f4(f7)) all have inherited this, yet in various versions, while F6(f4(f1)) has a deviating version (namely 'deselection' to promote survival of the fitted individuals). It seems extremely unlikely that all sub-functions under each of these functions will have the same principles and mechanisms and that this will be the case for the next fractal, and the next, and so on, because if so then the fractal principle would also imply that the previous fractal level had the same mechanisms, which is incorrect. So, it is highly likely that the sub-functions will have varying mechanisms, much more detailed modes of mechanisms and/or more specifically diverging mechanisms to fulfil the role of the individual sub-functions.

This may explain the puzzle in Section 9.3.1 in which a mechanism seem to contradict (relax) the mechanism of natural selection: deeper levels of detail obtained at a next fractal level of the model may provide an explanation.

Inspired by Koonin mentioning 'alternative splicing' under the heading of "Emergence and evolution of genomic complexity …" ([9] *page 1022*), the present author would like to emphasize the definition of 'evolution' in Section 1.2. Alternative splicing definitely is a major facet generating complexity, and the evolutionary mechanism that generated the alternative splicing has provided a mechanism that generates additional complexity into the variation that feeds evolutionary processes; however, this does not make alternative splicing an evolutionary theory.

### *18.5.2 Now, What about the Existing Evolutionary Theories?*

First, the Unifying Theory of Evolution does not discount or discard any existing theory. The main achievement has been to differentiate existing theories and their component elements and then integrating them in a hierarchical framework, however, this should not be misinterpreted as a value-based differential treatment. Each of these theories is placed in a shared framework where they represent cooperative theories. At the same time, the framework tells which of them operate in parallel and which operate in succession.

The exploration of existing theories may continue, in principle unaffected by the Unifying Theory of Evolution, as each such theory constitutes a piece in the puzzle instituting the wholeness of biological evolution.

### *18.5.3 What's the use of The Unifying Theory?*

As a consequence of the conclusion in the previous section, one major contribution that the new framework may provide is as follows: as existing theories mature or are exhausted they may be combined to gradually create a greater picture that may be analysed by gradually more advanced and integrated quantitative simulations. Without the framework, the alternative would be to combine theories in a trial and error fashion at advanced simulation studies. The three perspectives in the structure operate in parallel, while the functions operate in sequence – and hence, these work somewhat like Russian Dolls. Therefore, advanced simulation studies may exploit the framework in the Unifying Theory of Evolution to tell which theories may be combined and in which sequence.

Further, the framework explicitly points out areas for future research, in that some topics in the framework are weakly populated with theories and/or theories that are sparsely investigated.

Finally, this kind of modelling may constitute a new way of writing textbooks because of its logical and highly structured organisation of information, and because it brings things into a context rather than just bringing information purely topic oriented; the important is that the template information model through its structure is guiding the information search in an appropriate manner. However, for a university level textbook like [195] this may require a total of 4-5 fractal levels, including the macro-level – that is, a Sisyphean labour.

### *18.5.4 Competing or Conflicting Unifying Theories?*

What does it require to be called a unified or unifying theory? It requires that multiple theories that are previously treated as separate, but not necessarily mutually independent, are united into one functional wholeness. According to [5], 'unify' (unifying, unified) means "to make or become one; unite". The notion 'unifying' is preferred for the present theory as it denotes that a synthesis of this kind is an active and ongoing process and that it is not necessarily completed or closed in its first version.

A literature search for 'Unified Theory' or 'Unifying Theory' and 'evolution' (in the title or abstract) gave sparse results, of which only a couple were relevant or marginally relevant.

Koonin and Wolf in [24] (in their *Figure 8*) bring forward a united conceptual structure of evolutionary biology where they 'integrate' a series of paradigm evolutionary theories, from Lamarckian, over Darwin and the quantitative theory of selection and drift – used as a foundation for the rest: selfish genes, neutral theory, constructive neutral evolution of complexity horizontal gene transfer, phylogenomics, and neo-Lamarckian evolution models; these latter are placed almost in parallel, while evolution of evolvability is paralleled with the Darwinian theory of evolution. The conceptual structure constitutes a unified theory of evolution in that it is somewhat hierarchical with parallel tracks, all depicting the timewise progression of ideas from basic principles to more and more detailed principles, for instance in the path from Lamarck, over Darwin, to neutral theory and on to the constructive neutral evolution theory. The similarity with the Unified Theory of Evolution is the parallel tracks, while the difference is that the one has a chronological progression, while the other has a functional progression. There is no conflicting information between the two; for instance, horizontal gene transfer (a set of evolutionary mechanisms), phylogenomics (the study of evolutionary relationships between genes, populations, species, etc. ([11])), selfish genes (a set of mechanisms promoting specific kinds of evolutionary change) and evolution of evolvability (the analysis of evolutionary mechanisms for evolution itself) are all included in both. Of these latter topics, the horizontal gene transfer and the selfish genes are referred to Sections 9.2 and 9.3, simply because they are mechanisms creating genomic variation and therefore may contribute to the evolution of evolvability dealt with in the mentioned sections.

Frías' statement "The Synthetic Theory of Evolution is the most unifying theory of life science." ([130] *page 299*) as the very first statement of the abstract and of the Introduction naturally captured the interest. And he continues with "This theory is mainly based on Neo-Darwinism, particularly on Mendelism, population genetics, mutations, natural selection, gradualism, and the central dogma of molecular biology." ([130] *page 299*). Moreover, after reading the paper, a check on Wikipedia (last accessed on the 18th of May 2016) verified that this beloved child has many names: "The *modern evolutionary synthesis* (known as the *new synthesis*, the *modern synthesis*, the *evolutionary synthesis*, *millennium synthesis* or the *neo-Darwinian synthesis*) …". So, in reality, this corresponds to the Standard Evolutionary Theory. Frías does combine several smaller theories into one larger framework (see Table 1), as also mentioned in the second citation above. In essence, he expands the Neo-Darwinian thinking with horizontal gene transfer and selfish genes (see Section 9.2.1 and 9.3.1). Compared to the present Unifying theory, Frías' Synthetic Theory of Evolution is considered a component element – very much corresponding to the Standard Evolutionary Theory, but which in some areas digs a bit deeper and hence might serve as supplement for an extension into the next fractal level. His theory is neither competitive, nor conflicting with the present Unifying Theory of Evolution, except that horizontal gene transfer in the present study is divided on two perspectives rather than considered belonging entirely under neo-Darwinian thinking.

Skinner recently presented a different Unified Theory that describes the integration (or rather merge) of environmental epigenetic and genetic aspects of evolution; see [36]. This theory is neither competitive, nor contradicting the present Unifying Theory of evolution. Rather, he argues in favour of epigenetics as an evolutionary factor in its own right and describes how the traditional view on evolution (called Neo-Darwinian evolution, i.e. again the Standard Evolutionary Theory) play together with Neo-Lamarckian evolution – that is, the Extended Evolutionary Synthesis. In his proposal for a Unified Theory, the shared point between genetics-driven evolution and epigenetics' contribution to evolution is the 'Genome & DNA Sequence': "Environmental epigenetic alterations promote genetic mutations to alter genotypic variation. Environmental epigenetics and genetic mutations both promote phenotypic alterations on which natural selection acts" ([36] *Table 1 and Figure 1*). That is, he unites the two perspectives into a single wholeness. Seen from the viewpoint of the present Unifying Theory of Evolution, Skinners model shows the dynamic interplay between the two perspectives, Standard Evolutionary Theory and Extended Evolutionary Synthesis, respectively; and hence there is no conflict.

Sharma in [197] presents a 'Unified Theory of Biology' as suggested in the title of the paper. The author propose a step-wise model to explain "transgenerational epigenetic inheritance and its evolutionary significance by integrating gene expression and gene networks, miRNA or other RNA, DNA methylation, histone modifications, and DNA-methylation-induced mutation, in three mechanistic steps" ([197] *page 3368*). The first step is concerned with the communication of heritable information about environmental effects in somatic cells to the germ line by means of circulating RNAs (representing invironmental, physiological conditions). The second step comprises epigenetic modifications propagated across generations through gene expression and gene networks. Finally, the third step: "inherited epigenetic variations, represented by methylated cytosines, are fixed in the population as thymines in an evolutionary time scale." ([197] *page 3386*). There is neither competition nor any conflict between Sharma's Unified Theory of Biology and the present Unifying Theory of evolution; rather the former may serve in some specific points as an explanatory model at a more detailed molecular level than chosen for the present model, but it is not seen as a competing theory of evolution.

Sameroff in [198] propose a unified theory of development that integrates four models used for investigating individuals' psychological patterns and predicting their change over time: personal change, context, regulation and representational models of human personal development: The personal change model facilitates understanding the progression of competencies from infancy to adulthood; the contextual model delineates the multiple sources of experience that augment or constrain individual development; his regulation model provides a dynamic systems perspective to the previous two models; and his representational model express an individual's experiences in the world. Now, it is obvious that this unified model is connected with the Life-History Theory in Section 8.3.3.1 and may be applied to explore this further. Thus, there is no competition or conflict between this and the Unifying Theory of Evolution.

### *18.5.5 Lessons Learned*

The major lessons learned from this study are described in the following. These are one result from the reflexivity issue discussed in Section 18.2.2.

Much to the surprise of the author, very few of the reviews, which were considered relevant, are systematic reviews in the sense displayed by Hummel and co-workers in [157]. One problem causing this may be that there are established schools within evolutionary biology (e.g. Standard Evolution Theory versus Extended Evolutionary Synthesis as well as the Constructive Neutral Theory). A derived problem is the excluding nature of some reviews, borderlining a bias originating in perspectives; see also the discussion in the Foreword. The real problem arises when a review is subjective, because the emotional factor may enforce the problem pointed at in the Foreword. In other words one has to be extraordinary cautious of the entire article when meeting phrases like "such speculation ...", "this stance is not very different from the intelligent-design philosophy" and "a patently non-scientific irrational idea". Science must unconditionally stay objective in nature and in the accomplishment of a study, and phrases like the mentioned only act as fuel on an existing fire. Fortunately, given the nature of the Unifying Theory of Evaluation this is not unconditionally a problem in the present study: because of the structure with three perspectives such biased reviews tend to fit rather precisely into one of the perspectives, while articles of an objective nature tend to deal with topics horizontally (i.e. mixing the SET and the EES perspective). Nevertheless, – if not rejected for other reasons as sources of information – such articles have to be applied with extraordinary caution.

Another problem – at least initially for a reader – is the terminology; beloved child has many and conflicting names. For instance, plasticity has an adaptive nature, but the neo-Darwinian perspective on evolution is by some characterised as 'adaptive', while the neo-Lamarckian perspective that includes 'plasticity' and 'epigenetics' is characterised as 'non-adaptive'; see the discussion in Chapter 3. This shows that definitions are essential to avoid confusion and misinterpretations.

One has to be very careful about the context of statements given in source articles; this is particularly prominent in articles with a subjective twist, where sometimes the context is explicated in a subsequent or a preceding sentence. An example: "All replicating populations are capable of evolution, but it has recently been argued that some species are better at it than others, with natural selection directly advancing features of genomic architecture, genomic networks, and developmental pathways to promote the future ability of a species to adaptively evolve. Such speculation, which is almost entirely restricted to molecular and cell biologists and those who study digital organisms (e.g. *refs*) has been subject to considerable criticism by evolutionary biologists (e.g. *refs*).". First, the statement is presented as a complete sentence revealing a 'fact' seemingly elicited or synthesized from the literature, while in the subsequent sentence, the sarcasm reveals the emotional attitude of the author and his disbelief of such conclusion. Consequently, if using such article as 'evidence' then one has to be extremely careful with the context of every statement adopted.

# 19 References

[1] J. Brender McNair, P. McNair, L. Dennis, Z. Tümer, ATCG - An Applied Theory for Human MoleCular Genetics. In: Dennis L, Brender McNair J, Kauffman LH, eds., The Mereon Matrix: Unity, Perspective and Paradox, Elsevier, London, 2013.

[2] L. Dennis, J. Brender McNair, L.H. Kauffman, eds., The Mereon Matrix: Unity, Perspective and Paradox, Elsevier, London, 2013.

[3] K.N. Laland, T. Uller, M.W. Feldman, K. Sterelny, G.B. Müller, et al., The extended evolutionary synthesis: its structure, assumptions and predictions, *Proc Biol Sci* **282(1813)** (2015), 20151019. (available at http://www.ncbi.nlm.nih.gov/pmc/articles/PMC4632619/. Last accessed May 2016).

[4] K. Laland, T. Uller, M. Feldman, K. Sterelny, G.B. Müller, et al., Does evolutionary theory need a rethink? *Nature* **514(7521)** (2014), 161-164.

[5] *Collins English Dictionary, Third Updated Edition*. HarperCollins Publishers, Glasgow, 1995.

[6] L. von Bertalanffy, *General Systems Theory, Foundations, Development, Applications.* Revised edition (of the original 1968 edition, 17th paperback printing, 2009), George Braziller, New York, 1969.

[7] M. Kirschner, Beyond Darwin: evolvability and the generation of novelty, *BMC Biol* **11** (2013), 110. (available at http://www.biomedcentral.com/1741-7007/11/110. Last accessed June 2016).

[8] L. Dennis, R.W. Gray, L.H. Kauffman, J. Brender McNair, N.J. Woolf, A Framework Linking Non-Living and Living Systems: Classification of Persistence, Survival and Evolution Transitions, *Foundations of Science* **14(3)** (2009), 217-238.

[9] E.V. Koonin, Darwinian evolution in the light of genomics, *Nucleic Acids Res* **37(4)** (2009), 1011-1134.

[10] M. Lynch, The frailty of adaptive hypotheses for the origins of organismal complexity, *Proc Natl Acad Sci U S A* **104 Suppl 1** (2007), 8597-604. (Last accessed 16th of June 2016.Accessible at http://www.ncbi.nlm.nih.gov/pmc/articles/PMC1876435/.)

[11] R. Cammack, T.K. Attwood, P.N. Campbell, J.H. Parish, A.D. Smith, et al. (eds.), *Oxford Dictionary of Biochemistry and Molecular Biology, ed 2*, Oxford University Press, Oxford, 2006.

[12] F.C. Boogerd, F.J. Bruggeman, R.C. Richardson, Mechanistic explanations and models in molecular systems biology, *Found Sci* **18** (2013), 725-744.

[13] J. Brender, J. Talmon, On Using references as evidence, *Meth Inf Med* **48(6)** (2009), 503-507.

[14] L. Mathiassen, A. Munk-Madsen, Formalizations in systems development, *Behavior Inf Technol* **5(2)** (1986), 145-155.

[15] J. Brender, Handbook of Evaluation Methods for Health Informatics, Academic Press/Elsevier, New York, 2006.

[16] C. Hampden-Turner, F. Trompenaars, The seven cultures of capitalism, Doubleday, New York, 1993.

[17] C. Hampden-Turner, F. Trompenaars, Mastering the infinite game – how East Asian values are transforming business practices, Capstone Publishing Ltd, Oxford, 1997.

[18] Trompenaars, C. Hampden-Turner, Riding the waves of culture – understanding cultural diversity in business, 2nd ed. Nicholas Brealey Publishing Ltd, London, 1997.

[19] S.C. Schneider, J.-L. Barsoux, Managing across cultures, 2nd ed., FT Prentice Hall, Harlow, 2003.

[20] C. Darwin, On the Origin of Species by Means of Natural Selection, or the Preservation of Favoured Races in the Struggle for Life, 5th Edition, John Murray, London, 1869. (available at http://darwin-online.org.uk/converted/pdf/1869_Origin_F387.pdf. Last accessed January 2014).

[21] S.A. Frank, Natural selection. II. Developmental variability and evolutionary rate, *J Evol Biol* **24** (2011), 2310-2320.

[22] L. Witting, Inevitable evolution: back to The Origin and beyond the 20th century paradigm of contingent evolution by historical natural selection, *Biol Rev Camb Philos Soc* **83(3)** (2008), 259-294.

[23] A. Stoltzfus, Constructive neutral evolution: exploring evolutionary theory's curious disconnect, *Biol Direct* **7** (2012), 35, doi: 10.1186/1745-6150-7-35 (accessible from http://www.ncbi.nlm.nih.gov/pmc/articles/PMC3534586/pdf/1745-6150-7-35.pdf. Last accessed April 2016).

[24] E.V. Koonin, Y.I. Wolf, Evolution of microbes and viruses: a paradigm shift in evolutionary biology? *Front Cell Infect Microbiol* **Sep 13;2** (2012), 119. doi: 10.33/fcimb.2012.00119.

[25] X. Wang, C. Chen, L. Wang, D. Chen, W. Guang, J. French, Conception, early pregnancy loss, and time to clinical pregnancy: a population-based prospective study, *Fertil Steril* **79(3)** (2003), 577-584.

[26] T. Lenormand, D. Roze, F. Rousset, Stochasticity in evolution, *Trends Ecol Evol* **24(3)** (2009), 157-165.

[27] S.M. Scheiner, Bet-hedging as a complex interaction among developmental instability, environmental heterogeneity, dispersal, and life-history strategy, *Ecol Evol* **4(4)** (2014), 505-515.

[28] P. Bak, S. Boettcher, Self-organized criticality and punctuated equilibria, *Physica D* **107(2-4)** (1997), 143-150.

[29] R.V. Solé, S.C. Manrubia, M. Benton, S. Kauffman, P. Bak, Criticality and scaling in evolutionary ecology, *Trends Ecol Evol* **14(4)** (1999), 156-160.

[30] J. Hesse, T. Gross, Self-organized criticality as a fundamental property of neural systems, *Front Syst Neurosci* **8** (2014), article 166. doi: 10.3389/fnsys.2014.00166. (Last accessed 16th of June 2016. Accessible at http://journal.frontiersin.org/article/10.3389/fnsys.2014.00166/full).

[31] N.A. Bihlmeyer, J.A. Brody, A. Smith, K.L. Lunetta, M. Nalls, et al., Genetic diversity is a predictor of mortality in humans, *BMC Genet* **15(159)** (2014), 1-7.

[32] H. Innan, F. Kondrashov, The evolution of gene duplication: classifying and distinguishing between models, *Nat Rev Genet* **11(2)** (2010), 97-108.

[33] P. Beldade, A.R.A. Mateus, R.O. Keller. Evolution and molecular mechanisms of adaptive plasticity, *Mol Ecol* **20** (2011), 1347-1363.

[34] K.N. Laland, J. Odling-Smee, M.W. Feldman, J. Kendal, Conceptual barriers to progress within evolutionary biology, *Found Sci* **14(3)** (2009), 195-216.

[35] T. Uller, Evolutionary perspectives on transgenerational epigenetics. In: [38], 175-185.

[36] M.K. Skinner, Environmental epigenetics and a Unified Theory of molecular aspects of evolution: a Neo-Lamarckian concept that facilitates Neo-Darwinian evolution, Genome Biol Evol **7(5)** (2015), 1296-1302.

[37] H. Cedar, Y. Bergman, Programming of DNA methylation patterns, *Annu Rev Biochem* **81** (2012), 97-117.

[38] T. Tollefsbol, ed., Transgenerational Epigenetics: Evidence and Debate, Academic Press, Amsterdam, 2014.

[39] S. Valena, A.P. Moczek, Epigenetic mechanisms underlying developmental plasticity in horned beetles, *Genet Res Int* **2012** (2012), 576303. doi: 10.1155/2012/576303.

[40] N. Aubin-Horth, S.C. Renn, Genomic reaction norms: using integrative biology to understand molecular mechanisms of phenotypic plasticity, *Mol Ecol* **18(18)** (2009), 3763-3780.

[41] J. Flegr. Influence of latent *Toxoplasma* infection on human personality, physiology and morphology: pros and cons of the *Toxoplasma*-human model in studying the manipulation hypothesis. *J Exp Biol* **216** (2013), 127-133.

[42] J. Flegr, A. Markoš, Masterpiece of epigenetic engineering – how *Toxoplasma gondii* reprogrammes host brains to change fear to sexual attraction, *Mol Ecol* **23(24)** (2014), 5934-5936.

[43] T.B. Cook, L.A. Brenner, C.R. Cloninger, P. Langenberg, A. Igbide, et al., "Latent" infection with *Toxoplasma gondii*: association with trait aggression and impulsivity in healthy adults, *J Psychiatr Res* **60** (2015), 87-94.

[44] R. Kellermayer, Metastable Epialleles. In: [38], 69-73.

[45] T.A. Kosten, D.A. Nielsen, Maternal Epigenetic inheritance and stress during gestation: focus on brain and behavioural disorders. In: [38], 197-219.

[46] C. Giuliani, M.G. Bacalini, M. Sazzini, C. Pirazzini, C. Franceschi, et al., The epigenetic side of human adaptation: hypotheses, evidences and theories, *Ann Hum Biol* **42(1)** (2015), 1-9.

[47] D.W. Pfennig, M.A. Wund, E.C. Snell-Rood, T. Cruickshank, C.D. Schlichting, A.P. Moczek, Phenotypic plasticity's impact on diversification and speciation, *Trends Ecol Evol* **25** (2010), 459-467.

[48] C.D. Schlichting, M.A. Wund, Phenotypic plasticity and epigenetic marking: an assessment of evidence for genetic accommodation, *Evol* **68(3)** (2014), 656-672.

[49] A.A. Sharov, Evolutionary constraints or opportunities? *BioSystems* **123** (2014), 9-18.

[50] E.J. Ragsdale, M.R. Müller, C. Rödelsperger, R.J. Sommer, A developmental switch coupled to the evolution of plasticity acts through a sulfatase, *Cell* **155** (2013), 922-933.

[51] I.G. Romero, I. Ruvinsky, Y. Gilad, Comparative studies of gene expression and the evolution of gene regulation, *Nat Rev Genet* **13(7)** (2012), 505-16.

[52] D. Tautz. Evolution of transcriptional regulation, *Curr Opin Genet Dev* **10(5)** (2000), 575-579.

[53] P. Landi, C. Hui, U. Dieckmann. Fisheries-induced disruptive selection, *J Theor Biol* **365** (2015), 204-216.

[54] A.P. Moczek, Developmental capacitance, genetic accommodation, and adaptive evolution, *Evol Dev* **9(3)** (2007), 299-305.

[55] A.P. Moczek, S. Sultan, S. Foster, C. Ledón-Rettig, I. Dworkin, et al., The role of developmental plasticity in evolutionary innovation, *Proc R Soc B* **278** (2011), 2705-2713.

[56] E.C. Snell-Rood, J.D. Van Dyken, T. Cruickshank, M.J. Wade, A.P. Moczek, Toward a population genetic framework of developmental evolution: the costs, limits, and consequences of phenotypic plasticity, *BioEssay* **32** (2010), 71-81.

[57] T. Schwander, R. Libbrecht, L. Keller, Supergenes and complex phenotypes, *Curr Biol* **24(7)** (2014), R288-R294.

[58] M.J. Thompson, C.D. Jiggins, Supergenes and their role in evolution, *Heredity* **113(1)** (2014), 1–8.

[59] D. Melo, G. Marroig. Directional selection can drive the evolution of modularity in complex traits, *PNAS* **112(2)** (2015), 470-475.

[60] S.A. Frank, Natural selection. VII. History and interpretation of kin selection theory, *J Evol Biol* **26** (2013), 1151-1184.

[61] A. Ross-Gillespie, R. Kümmerli, Collective decision-making in microbes, *Front Microbiol* **5** (2014), article 54. doi: 10.3389/fmicb.2014.00054. (Last accessed 16th of June 2016. Accessible at http://journal.frontiersin.org/article/10.3389/fmicb.2014.00054/full).

[62] A.A. Dandekar, S. Chugani, E.P. Greenberg, Bacterial quorum sensing and metabolic incentives to cooperate, *Science* **338** (2012), 264-266.

[63] A. Prindle, J. Liu, A. Asally, S. Ly, J. Garcia-Ojalvo, G.M. Süel, Ion channels enable electrical communication in bacterial communities, *Nature* **527** (2015) 59-63.

[64] T.G. Benton, S.J. Plaistow, I.N. Coulson, Complex population dynamics and complex causation: devils, details and demography, *Proc R Soc B* **273** (2006), 1173-1181.

[65] S. Kapeller, H. Schroeder, S. Schueler, Modelling the spatial population dynamics of the green oak leaf roller (*Tortrix viridana*) using density dependent competitive interactions: Effect of herbivore mortality and varying host-plant quality, *Ecol Modell* **222** (2011), 1293-1302.

[66] M.M. Webster, A.J.W. Ward, Personality and social context, *Biol Rev Camb Philos Soc* **86** (2011), 759-775.

[67] M.A. Nowak, Five rules for the evolution of cooperation, *Science* **314** (2006), 1560-1563.

[68] P. Abbot, J. Abe, J. Alcock, S. Alizon, J.A.C. Alpedrinha, et al., Inclusive fitness theory and eusociality, *Nature* **471 (7339)** (2011), E1-4.

[69] G.A.K. Wyatt, S.A. West, A. Gardner, Can natural selection favour altruism between species? *J Evol Biol* **26** (2013), 1854-1865.

[70] J.D. van Dyken, M.J. Wade, Origins of altruism diversity I: The diverse ecological roles of altruistic strategies and their evolutionary responses to local competition, *Evol* **66-8** (2012 a), 2484-2497.

[71] J.D. van Dyken, M.J Wade, Origins of altruism diversity II: Runaway coevolution of altruistic strategies via "Reciprocal Niche Construction", *Evol* **66(8)** (2012 b), 2498-2513.

[72] L. Lehmann, L. Keller, The evolution of cooperation and altruism – a general framework and a classification of models, *J Evol Biol* **19(5)** (2006), 1365-1376.

[73] P.E. Smaldino, The cultural evolution of emergent group-level traits, *Behav Brain Sci* **37(3)** (2014), 243-254.

[74] R. Bergmüller, R. Schürch, I.M. Hamilton, Evolutionary causes and consequences of consistent individual variation in cooperative behaviour, *Philos Trans R Soc Lond B Biol Sci* **365(1553)** (2010), 2751-2764. doi: 10.1098/rstb.2010.0124. (Last accessed 16th of June 2015. Available from: http://rspb.royalsocietypublishing.org/.).

[75] S.A. West, A. Gardner, Altruism, spite, and greenbeards, *Science* **327** (2010), 1341-1344.

[76] G. Hein, Y. Morishima, S. Leiberg, S. Sul, E. Fehr, The brain's functional network architecture reveals human motives, *Science* **351(6277)** (2016), 1074-1078.

[77] N.M. Marples, D.J. Kelly, R.T. Thomas, Perspective: the evolution of warning coloration is not paradoxical, *Evol* **59(5)** (2005), 933-940.

[78] E.V. Koonin, Y.I.Wolf, Just how Lamarckian is CRISPR-Cas immunity: the continuum of evolvability mechanisms, *Biol Direct* **11(1)** (2016), 9. doi: 10.1186/s13062-016-0111-z. (available from http://www.ncbi.nlm.nih.gov/pmc/articles/PMC4765028/. Last accessed June 2016).

[79] M. Wolf, F.J. Weissing, An explanatory framework for adaptive personality differences, *Philos Trans R Soc Lond B Biol Sci* **365(1560)** (2010), 3959-3968.

[80] M. Briffa, J. Greenaway. High In Situ repeatability of behaviour indicates animal personality in the beadlet anemone Actinia equine (Cnidaria), *PLoS ONE* **6(7)** (2011), e21963. (available at http://www.ncbi.nlm.nih.gov/pmc/articles/PMC3130786/pdf/pone.0021963.pdf. Last accessed 16th of June 2016).

[81] N.L. Foster, M. Briffa, Familial strife on the seashore: aggression increases with relatedness in the sea anemone Actinia equine, *Behav Processes* **103** (2014), 243-245.

[82] I. Planas-Sitjà, J.L. Deneubourg, C. Gibon, G. Sempo, Group personality during collective decision-making: a multi-level approach, *Proc R Soc B* **282** (2015), 20142515. doi: 10.1098 / rspb.2014.2515. (Last accessed 16th of June 2016. accessible via http://rspb.royalsocietypublishing.org/content/royprsb/282/1802/20142515.full.pdf).

[83] M. Wolf, G. Sander Van Dorn, F.J. Weissing, On the coevolution of social responsiveness and behavioural consistency, *Proc R Soc B* **278** (2011), 440-448.

[84] E.N. Aron, A. Aron, J. Jagiellowicz, Sensory processing sensitivity: a review in the light of the evolution of biological responsivity, *Pers Soc Psychol Rev* **16(3)** (2012), 262-282.

[85] S. Luo, A unifying framework reveals key properties of multilevel selection, *J Theor Biol* **341** (2014), 41-52.

[86] T. Pievani, Individuals and groups in evolution: Darwinian pluralism and the multilevel selection debate, *J Biosci* **39(2)** (2014), 319-325.

[87] D. Gerkey, L. Cronk, What is a group? Conceptual clarity can help integrate evolutionary and social scientific research on cooperation, *Behav Brain Sci* **37(3)** (2014), 260-261.

[88] A. Gardner. The genetical theory of multilevel selection, *J Evol Biol* **28** (2015), 305-319.

[89] F. Débarre, Fitness costs in spatially structured environments, *Evol* **69(5)** (2015), 1329-1335.

[90] J. Krause, R.K. Butlin, N. Peuhkuri, V.L. Pritchard, The social organization of fish shoals: a test of the predictive power of laboratory experiments for the field, *Biol Rev Camb Philos Soc* **75(4)** (2000), 477-501.

[91] A. Cavagna, I. Giardina, T.S. Grigera, A. Jelic, D. Levine, S. Ramaswamy, M. Viale. Silent flocks: constraints on signal propagation across biological groups, *Phys Rev Lett* **114(21)** (2015), 218101. (epub, DOI: 10.1103/PhysRevLett.114.218101. Last accessed 16th of June 2016).

[92] M. van Veelen, J. García, M.W. Sabelis, M. Egas, Group selection and inclusive fitness are *not* equivalent; the Price equation vs. models and statistics, *J Theor Biol* **299** (2012), 64-80.

[93] L. Lehmann, L. Keller, West S, Roze D, Group selection and kin selection: Two concepts but one process, *Proc Natl Acad Sci* **104(16)** (2007), 6736-6739.

[94] M. van Veelen, Group selection, kin selection, altruism and cooperation: When inclusive fitness is right and when it can be wrong, *J Theor Biol* **259** (2009), 589-600.

[95] B. Simon, J.A. Fletcher, M. Doebeli, Towards a general theory of group selection, *Evol* **67(6)** (2013), 1561-1572.

[96] M. van Veelen, S. Luo, B. Simon, A simple model of group selection that cannot be analysed with inclusive fitness, *J Theor Biol* **360** (2014), 279-289.

[97] B. Allen, M.A. Nowak, E.O. Wilson, Limitations of inclusive fitness, *PNAS* **110(50)** (2013), 20135-20139.

[98] M. Ghoul, A.S. Griffin, S.A. West, Toward and evolutionary definition of cheating, *Evol* **68(2)** (2013), 318-331.

[99] E.A. Ostrowski, Y. Shen, X. Tian, R. Susgang, H. Jiang, et al., Genomic signatures of cooperation and conflict in the social amoeba, *Curr Biol* **25** (2015), 1661-1665.

[100] G.J. Velicer, Evolution of cooperation: does selfishness restraint lie within? *Curr Biol* **15(5)** (2005), R173-R175.

[101] A.L. Cronin, Ratio-dependent quantity discrimination in quorum sensing ants, *Anim Cogn* **17** (2014), 1261-1268.

[102] R. Jeanson, A. Dussutour, V. Fourcassié. Key factors for the emergence of collective decision in invertebrates, *Front Neurosci* **6** (2012), 121. (Last accessed 16th 0f June 2016. Available at http://journal.frontiersin.org/article/10.3389/fnins.2012.00121/full.).

[103] I.D. Couzin, Collective cognition in animal groups, *Trends Cogn Sci* **13(1)** (2009), 36-43.

[104] J. Cote, J. Clobert, T. Brodin, S. Fogarty, A. Sih, Personality-dependent dispersal: characterisation, ontogeny and consequences for spatially structured populations, *Phil Trans R Soc B* **365** (2010), 4065-4076.

[105] M. Pelé, C. Sueur, Decision-making theories: linking the disparate research areas of individual and collective cognition, *Anim Cogn* **16** (2013), 543-556.

[106] J. Clobert, J.-F. Le Galliard, J. Cote, S. Meylan, M. Massot, Informed dispersal, heterogeneity in animal dispersal syndromes and the dynamics of spatially structured populations, *Ecol Lett* **12(3)** (2009), 197-209.

[107] D.E. Bowler, T.G. Benton, Causes and consequences of animal dispersal strategies: relating individual behaviour to spatial dynamics, *Biol Rev* **80** (2005), 205-225.

[108] C.J. Ingram, C.A. Mulcare, Y. Itan, M.G. Thomas, D.M. Swallow, Lactose digestion and the evolutionary genetics of lactase persistence, *Hum Genet* **124(6)** (2009), 579-591. As a fundamental property of neural systems.

[109] M.E. Allentoft, M. Sikora, K.G. Sjögren, S. Rasmussen, M. Rasmussen, et al., Population genomics of Bronze Age Eurasia, *Nature* **522(7555)** (2015), 167-172.

[110] L. Conradt, Models in animal collective decision-making: Information uncertainty and conflicting preferences, *Interface Focus* **2** (2012), 226-240.

[111] O. Petit, R. Bon, Decision-making processes: the case of collective movements, *Behav Processes* **84** (2010), 635-647.

[112] L.M. Aplin, D.R. Farine, R.P. Mann, B.C. Sheldon, Individual-level personality influences social foraging and collective behaviour in wild birds, *Proc R Soc B* **281** (2014), 20141016. (available from http://rspb.royalsocietypublishing.org/. Last accessed 16th June 2016).

[113] G. Kerth, Group decision-making in fission-fusion societies, *Behav Processes* **84(3)** (2010), 662-663.

[114] J.A. Merkle, M. Sigaud, D. Fortin. To follow or not? How animals in fission-fusion societies handle conflicting information during group decision-making, *Ecol Lett* **18(8)** (2015), 799-806.

[115] D. Fleischmann, I.O. Baumgartner, M. Erasmy, N. Gries, M. Melber, et al., Female Bechstein's bats adjust their group decisions about communal roosts to the level of conflict of interests, *Curr Biol* **23(17)** (2013), 1658-1662.

[116] P. Richerson, R. Baldini, A. Bell, K. Demps, K. Frost, et al., Cultural group Selection plays an essential role in explaining human cooperation: A sketch of the evidence, *Behavioral Brain Sci Behav Brain Sci* **39** (2016), 1-71. (originally viewed as the 'Epub ahead of print' from 2014 on http://journals.cambridge.org/action/displayAbstract?fromPage=online&aid=9395197&fileId=S01 ; printed version accessed 16th of June 2016).

[117] G. Hofstede, Cultures and Organizations, Software of the Mind, Intercultural Cooperation and its Importance for Survival, McGraw-Hill, New York, 1997.

[118] Á. Cabrera, E.F. Cabrera, S. Barajas, The key role of organizational culture in a multi-system view of technology-driven change, *Int J Inf Manag* **21** (2001), 245-261.

[119] T. Davis, E. Margolis, The priority of the individual in cultural inheritance, *Behav Brain Sci* **37(3)** (2014), 257-258.

[120] C. Rueffler, J. Hermisson, G.P. Wagner, Evolution of functional specialisation and division of labor. *PNAS* **109(6)** (2012), E326-E335.

[121] R. Libbrecht, L. Keller, The making of eusociality: insights from two bumblebee genomes, *Genome Biol* **16** (2015), 75. doi: 10.1186/s13059-015-0635-z (Last accessed 16th of June 2016. Available at http://www.ncbi.nlm.nih.gov/pmc/articles/PMC4408596/).

[122] J.L. Sachs, U.G. Mueller, T.P. Wilcox, J.J. Bull, The evolution of cooperation, *Q Rev Biol* **79(2)** (2004), 135-160.

[123] I. Barber, N.J. Dingemanse, Parasitism and the evolutionary ecology of animal personality, *Phil Trans R Soc B* **365** (2010), 4077-4088.

[124] E. Donadio, S.W. Buskirk, Diet, morphology, and interspecific killing in carnovora, *Am Nat* **167(4)** (2006), 524-36.

[125] J.L. Hoogland, C.R. Brown, Prairie dogs increase fitness by killing interspecific competitors, *Proc Biol Sci* **283(1827)** (2016). doi: 10.1098/rspb.2016.0144.

[126] B. Ekstig, Complexity, natural selection and the evolution of life and humans, *Found Sci* **20** (2015), 175-187.

[127] M. Sorek, E.M. Díaz-Almeyda, M. Medina, O. Levy, Circadian clocks in symbiotic corals: the duet between *Symbiodinium* algae and their coral host, *Mar Genomics* **14** (2014), 47-57.

[128] N.R. Franks, J.P. Stuttard, C. Doran, J.C. Esposito, M.C. Master, et al., How ants use quorum sensing to estimate the average quality of a fluctuating resource, *SCI Rep* **5** (2015), 11890; doi: 10.1038/srep11890 (2015). (accessible from: http://www.ncbi.nlm.nih.gov/pmc/articles/ PMC4495386/. Last accessed 16th June 2016).

[129] O. Seehausen, R.K. Butlin, I. Keller, C.E. Wagner, J.W. Boughman, et al., Genomics and the origin of species, *Nat Rev Genet* **15(3)** (2014), 176-192.

[130] D. Frías L., Omissions in the synthetic theory of evolution, *Biol Res* **43** (2010), 299-306.

[131] F.L. Mendez, G.D. Poznik, S. Castellano, C.D.Bustamante, The Divergence of Neandertal and Modern Human Y Chromosomes, *Am J Hum Genet* **98(4) (**2016),728-34.

[132] M.R. Kronforst, M.E.B. Hansen, N.G. Crawford, J.R. Gallant, W. Zhang, et al., Hybridization reveals the evolving genomic architecture of speciation, *Cell Reports* **6** (2013), 666-677.

[133] D.T. Iskandar, B.J. Evans, J.A. McGuire, A novel reproductive mode in frogs: A new species of fanged frog with internal fertilization and birth of tadpoles, *PLoS ONE* **9(12)** (2014), e115884. doi: 10.1371/journal.pone.0115884. (Last accessed 16th of June 2016. Available at http://www.ncbi.nlm.nih.gov/pmc/articles/PMC4281041/pdf/pone.0115884.pdf).

[134] S.F. Gilbert, T.C. Bosch, C.Ledón-Rettig, Eco-Evo-Devo: developmental symbiosis and developmental plasticity as evolutionary agents, *Nat Rev Genet* **16(10)** (2015), 611-622.

[135] L. Lehmann, V. Ravigne, L. Keller, Population viscosity can promote the evolution of altruistic sterile helpers and eusociality, *Proc B Soc B* **275** (2008), 1887-1895. (Last accessed 16th of June 2016. Available at http://rspb.royalsocietypublishing.org/.).

[136] T. Ferenci, R. Maharjan, Mutational heterogeneity: a key ingredient to bet-hedging and evolutionary divergence? *Bioessays* **37** (2014), 123-130.

[137] D.F. Simola, L. Wissler, G. Donahue, R.M. Waterhouse, M. Helmkampf, et al., Social insect genomes exhibit dramatic evolution in gene composition and regulation while preserving regulatory features linked to sociality, *Genome Res* **23(8)** (2013), 1235-1247.

[138] J.H. Hunt, A conceptual model for the origin of worker behaviour and adaptation of eusociality. *J Evol Biol* **25** (2012), 1-19.

[139] M.O. Krasnec, M.D. Breed, Eusocial evolution and the recognition systems in social insects, *Adv Exp Med Biol* **739** (2012), 78-92.

[140] P. Ibbotson, Group-level expression encoded in the individual, *Behav Brain Sci* **37(3)** (2014), 261-262.

[141] M.A. Nowak, C.E. Tarnitag, E.O. Wilson, The evolution of eusociality, *Nature* **466** (2010), 1057-1062.

[142] R. Libbrecht, P.R. Oxley, D.J.C. Kronauer, L. Keller, Ant genomics shes light on the molecular regulation of social organization, *Genome Biol* **14** (2014), 212. doi: 10.1186/gb-2013-14-7-212. (available at http://www.ncbi.nlm.nih.gov/pmc/articles/PMC4053786/. Last accessed 16th of June 2016).

[143] J.M. Jandt, S. Bengston, N. Pinter-Wollman, J.N. Pruitt, N.E. Raine, A. Dornhaus, A. Sih, Behavioural syndromes and social insects: personality at multiple levels, *Biol Rev Camb Philos Soc* **89(1)** (2014), 48-67.

[144] R. Jeanson, A. Weidenmüller, Interindividual variability in social insects - proximate causes and ultimate consequences, *Biol Rev Camb Philos Soc* **89(3)** (2014), 671-687. doi: 10.1111/brv.12074

[145] R.F.Oliveira, Social behavior in context: Hormonal modulation of behavioral plasticity and social competence, *Integr Comp Biol* **49(4)** (2009), 423-440.

[146] R.F.Oliveira, Social plasticity in fish: integrating mechanisms and function, *J Fish Biol* **81(7)** (2012), 2127-2150.

[147] S.M. Smith, T.E. Nichols, D. Vidaurre, A.M. Winkler, T.E. Behrens, et al., A positive-negative mode of population covariation links brain connectivity, demographics and behaviour, *Nat Neurosci* **18(11)** (2015), 1565-7.

[148] B.J. Ellis, E.A. Shirtcliff, W.T. Boyce, J. Deardorff, M.J. Essex, Quality of early family relationships and the timing and tempo of puberty: effects on biological sensitivity to context, *Dev Psychopathol* **23(1)** (2011), 85-99.

[149] B.H. Brumbach, A.J. Figuedo, B.J. Ellis, Effects of harsh and unpredictable environments in adolescence on development of life history strategies, *Hum Nat* **20** (2009), 25-51.

[150] J. Belsky, G.L. Schlomer, B.J. Ellis, Beyond cumulative risk: distinguishing harshness and unpredictability as determinants of parenting and early Life History Strategy, *Dev Psychol* **48(3)** (2012), 662-673.

[151] J.A. Simpson, V. Griskevicius, S.I.-C. Kuo, S. Sung, W.A. Collins, Evolution, stress, and sensitive periods: the influence of unpredictability in early versus late childhood on sex and risky behaviour, *Dev Psychol* **48(3)** (2012), 674-686.

[152] B.J. Ellis, The hypothalamic-pituitary-gonadal axis: a switch-controlled, condition-sensitive system in the regulation of life history strategies, *Horm Behav* **64(2)** (2013), 215-225.

[153] B.L. Brand, R.A. Lanius, Chronic complex dissociative disorders and borderline personality disorder: disorders of emotion dysregulation? *Borderline Personal Disord Emot Dysregul* **1** (2014), 13. doi: 10.1186/2051-6673-1-13. (Last accessed 11th April 2016. Accessible from http://www.bppded.com/content/1/1/13.).

[154] J.D. Ford, C.A. Courtois, Complex PTSD, affect dysregulation, and borderline personality disorder, *Borderline Personal Disord Emot Dysregul* **1** (2014), 9. doi: 10.1186/2051-6673-1-9. (accessible from http://www.bppded.com/content/1/1/9. Last accessed 11th April 2016).

[155] D. Mosquera, A. Gonzalez, A.M. Leeds, Early experience, structural dissociation, and emotional dysregulation in borderline personality disorder: the role of insecure and disorganized attachment, *Borderline Personal Disord Emot Dysregul* **1** (2014),15. doi: 10.1186/2051-6673-1-15. (accessible from http://www.bppded.com/content/1/1/135. Last accessed 11th April 2016).

[156] I.J. Rickard, W.E. Frankenhuis, D. Nettle, Why are childhood family factors associated with timing of maturation? A role for internal prediction, *Perspect Psychol Sci* **9(1)** (2014), 3-15.

[157] A. Hummel, K.H. Shelton, J. Heron, L. Moore, M.B.M. van den Bree, A systematic review of the relationships between family functioning, pubertal timing and adolescent substance use, *Addiction* **108** (2012), 487-496.

[158] D. Nettle, W.E. Frankenhuis, I.J. Rickard, The evolution of predictive adaptive responses in human life history, *Proc R Soc B* **280** (2013), 20131343 (Last accessed 16th of June 2016. Available from http://www.ncbi.nlm.nih.gov/pmc/articles/PMC3730599/).

[159] G. Davidowitz, H.F. Nijhout, D.A. Roff, Predicting the response to simultaneous selection: genetic architecture and physiological constraints, *Evol* **66(9)** (2012), 2916-2928.

[160] D.A. Roff, Contributions of genomics to life-history theory, *Nat Rev Genet* **8(2)** (2007), 116-25.

[161] E.J.S. Sonuga-Barke, Temporal discounting in conduct disorder: toward an experience-adaptation hypothesis of the role of psychosocial insecurity, *J Pers Disord* **28(1)** (2014), 19-24.

[162] M. Rutter, R. Kumsta, W. Schlotz, E. Sonuga-Barke, Longitudinal studies using a "natural experiment" design: the case of adoptees from Romanian institutions, *J Am Acad Child Adolesc Psychiatry* **51(8)** (2012), 762-770.

[163] R.A. Hoksbergen, J. ter Laak, C. van Dijkum, S. Rijk, K. Rijk, F. Stoutjesdijk, Posttraumatic stress disorder in adopted children from Romania, *Am J Orthopsychiatry* **73(3)** (2003), 255-265.

[164] M. Kennedy, J. Kreppner, N. Knights, R. Kumsta, B. Maughan, et al., Early severe institutional deprivation is associated with a persistent variant of adult attention-deficit/hyperactivity disorder: clinical presentation, developmental continuities and life circumstances in the English and Romanian Adoptees study, *J Child Psychol Psychiatry* (2016), 1-3. doi: 10.1111/jcpp.12576. (Epub ahead of print. Available from http://onlinelibrary.wiley.com/doi/10.1111/jcpp.12576/epdf. Last accessed 14th July 2016).

[165] V. Groza, S.D. Ryan, S.J. Cash, Institutionalization, behaviour and international adoption: predictors of behaviour problems, *J Immigr Health* **5(1)** (2003), 5-17.

[166] J. Tung, E.A. Archie, J. Altmann, S.C. Alberts, Cumulative early life adversity predicts longevity in wild baboons, *Nat Commun* **7** (2016):11181.

[167] W.W. Burggren, Dynamics of epigenetic phenomena: intergenerational and intragenerational phenotype 'washout', *J Exp Biol* **218(Pt 1)** (2015), 80-87.

[168] Y.-H. Jiang, J. Bressler, A.L. Beaudet, Epigenetics and human disease, *Ann Rev Genomics Hum Genet* **5** (2004), 479-510.

[169] L.J. Brent, D.W. Franks, E.A. Foster, K.C. Balcomb, M.A. Cant, et al., Ecological knowledge, leadership, and the evolution of menopause in killer whales, *Curr Biol* **25(6)** (2015), 746-750.

[170] K. Nishikawa, A.R. Kinjo, Cooperation between phenotypic plasticity and genetic mutations can account for the cumulative selection in evolution, *Biophysics* **10** (2014), 99-108.

[171] A.T. Fields, K.A. Feldheim, G.R. Poulakis, D.D. Chapman, Facultative parthenogenesis in a critically endangered wild vertebrate, *Curr Biol* **25** (2015), R446-R447.

[172] M. Loreau, C. de Mazancourt, Biodiversity and ecosystem stability: a synthesis of underlying mechanisms, *Ecol Ltt* **16** (2013), 106-115.

[173] W.L. Smith, J.H. Stern, M.G. Girard, M.P. Davis, Evolution of Venomous Cartilaginous and Ray-Finned Fishes, Integr Comp Biol (2016), doi: 10.1093/icb/icw070. (Epub ahead of print. Available from http://icb.oxfordjournals.org/content/early/2016/06/03/icb.icw070.full.pdf+html. Last accessed 14th July 2016).

[174] A.D. Foote, Y. Liu, G.W.C. Thomas, T. Vinař, J. Alföldi, et al., Convergent evolution of the genomes of marine mammals, *Nat Genet* **47(3)** (2015), 272-275.

[175] H.F. Nijhout, Big or fast: two strategies in the developmental control of body size, *BMC Biol* **13** (2015), 57. doi 10.1186/s12915-015-0173-x

[176] M. Doebeli, I. Ispolatov. Chaos and unpredictability in evolution, *Evol* **68(5)** (2014), 1365-1373.

[177] J. Draghi, G.P. Wagner. Evolution of evolvability in a developmental model, *Evol* **62(2)** (2007), 301-315.

[178] W.F. Doolittle, J. Lukeš, J.M. Archibald, P.J. Keeling, M.W. Gray, Comment on "Does constructive neutral evolution play an important role in the origin of cellular complexity?" *Bioessays* **33(6)** (2011),427-429.

[179] J. Lukeš, J.M. Archibald, P.J. Keeling, W.F. Doolittle, M.W. Gray, How a neutral evolutionary ratchet can build cellular complexity, *IUBMB Life* **63(7)** (2011), 528-537.

[180] D. Speijer, Does constructive neutral evolution play an important role in the origin of cellular complexity? Making sense of the origins and uses of biological complexity, *Bioessays* **33(5)** (2011), 344-349.

[181] W.F. Doolittle, Evolutionary biology: A ratchet for protein complexity, *Nature* **481(7381)** (2012), 270-271.

[182] H. Qiu, G. Cai, J. Luo, D. Bhattacharya, N. Zhang, Extensive horizontal gene transfers between plant pathogenic fungi, *BMC Biol* **14(1)** (2016), 41. (Last accessed 20th of June 2016. Accessible at http://bmcbiol.biomedcentral.com/articles/10.1186/s12915-016-0264-3.

[183] L. Gabora. Self-other organization: Why early life did not evolve through natural selection, *J Theor Biol* **241** (2006), 443-450.

[184] N.J. Woolf, L. Dennis, The Origin of Matter: Life, Learning and Survival. In: L. Dennis, J. Brender McNair, L.H. Kauffman, eds., The Mereon Matrix: Unity, Perspective and Paradox, Elsevier, London, 2013. pp. 305-345.

[185] A.V. Melkikh. Quantum information and the problem of mechanisms of biological evolution, *BioSystems* **115** (2014), 33-45.

[186] B. Vernot, J.M. Akey, Resurrecting Surviving Neandertal Lineages from Modern Human Genomes, *Science* **343** (2014), 1017-1021.

[187] L. Ermini, C. der Sarkissian, E. Willerslev, L. Orlando, Major transitions in human evolution revisited: A tribute to ancient DNA. J Human Evol **79** (2015), 4-20.

[188] J. Brender McNair. Theoretical basis of health IT evaluation. In: Ammenwerth E, Rigby M, eds. IOS Press, *Stud Health Technol Inform* **222** (2016), 38-52.

[189] J.C. Alderson, C. Clapham, D. Wall, Language test construction and evaluation, University Press, Cambridge 1995.

[190] J.H. Brown, V.K. Gupta, B.-L. Li, B.T. Milne, C. Restrepo, et al., The fractal nature of nature: power laws, ecological complexity and biodiversity, *Phil Trans R Soc Lond B* **357** (2002), 619-626.

[191] J.M. Fraser, P.J. Smith, A catalogue of errors, *Int J Man-Machine Stud* **37** (1992), 265-307.

[192] [No authors listed] Let's think about cognitive bias. *Nature* **526(7572)** (2015), 163.

[193] D. Jootun, G. McGhee, G.R. Marland, Reflexivity: promoting rigour in qualitative research, *Nurs Stand* **23(23)** (2009), 42-46.

[194] R. MacCoun, S. Perlmutter, Blind analysis: Hide results to seek the truth, *Nature* **526(7572)** (2015), 187-189.

[195] D.L. Nelson, M.M. Cox, *Lehninger, Principles of Biochemistry*, ed. 5, New York, WH Freeman and Company, 2008.

[196] M. Pavlicev, K. Hiratsuka, K.A. Swaggart, C. Dunn, L. Muglia, Detecting endogenous retrovirus-driven tissue-specific gene transcription, *Genome Biol Evol* **7(4)** (2015), 1082-1097

[197] A. Sharma, Systems genomics analysis centered on epigenetic inheritance supports development of a unified theory of biology, *J Exp Biol* **218** (2015), 3368-3373.

[198] A. Sameroff, A Unified Theory of Development: a dialectic integration of nature and nurture, *Child Dev* **81(1)** (2010), 6-22.

# Index

# E

# T



# U



# V



# W



# X

# Y

# Z